Daughter of the Dust

Tales from India & Africa

Marilyn S. Dodge

OLD AFRICA
STORIES *from* EAST AFRICA'S PAST

Published by Old Africa Books

Contents

"I do not want my home to be walled in on all sides and my windows to be stuffed. I want the cultures of all lands to be blown about my house as freely as possible."

—MAHATMA GANDHI

Acknowledgments

"We spend our years as a tale that is told . . . To everything there is a season."

—Ecclesiastes

Without the often unusual circumstances and exceptional people who have supported and walked the journey of life with me, there would be no substance for *Daughter of the Dust.*

Cole, my love, companion, best friend, critic and memory for much of this book, has encouraged and supported me in every endeavor I have pursued.

Our children, with their unwavering faith in the various decisions we made, added an element of exuberance I never thought possible under some of the conditions we lived with.

With comments such as "parts of a spectacular body searching for assembly" and "a rich story worth telling but not yet clear of the ground," my brothers John and Paul motivated me to organize my manuscript and add more detail.

Finally, this book was made a reality by the excellent editing and layout by Keith Miller.

Herewith, I offer the reader an often emotional story of our lives lived to the utmost of our abilities and to the glory of God.

If it's adventure you seek in this digital age
Look to Burton and Speke then turn the page.
We'll attempt to extol the heroic life
Of a man named Cole and his talented wife.
In mountainous lands both east and west,
And desert sands they stood the test.
There were endless trials and the poisoned dish,
Huge crocodiles and man-eating fish,
Death-dealing flus and confrontations,
Frightening coups and assassinations.

There's a passage herein for the ardent soul:
You don't always win when you're on a roll.
Though intrepid, unshaken and circumspect,
You sometimes get what you least suspect.
Cole's nerve never failed when the lions roared
But unfortunately he was nailed by a bungee cord!

—John C. Wiebe

Foreword

by Paul D. Wiebe

MARILYN IS MY younger sister. She and I have five older siblings. The seven of us were born and reared in south India by Mennonite missionary parents and grandparents.

Marilyn's and my parents and grandparents carried with them to India the good news of the gospel. They also built schools and clinics and hospitals, cared for orphans and the destitute, developed programs in adult education and literacy and other such endeavors, and encouraged the pursuit of justice. They called those who joined them in the Church, whatever their backgrounds, "brothers" and "sisters," in a land where the delineation of such relationships was very carefully constrained. However much they fell short, they did what they could to build up what was honest and true and of good report.

"Home" for our parents and grandparents and the seven of us children was some remembered—or, for those of us who had not yet been there, talked about—place far away. Home was also, always, wherever we were. We traveled and settled well as a family. We traveled and settled well together.

Marilyn grew up between social and cultural worlds, and her life since has been more, ever more, of the same. She married Cole Dodge, with whom she served in the Peace Corps in India, in 1966. Since then, she and Cole have lived and studied in Seattle and Berkeley; served with the American Friends Service Committee in Boston; directed the Quaker Peace and Service rehabilitation program in Nigeria following the Biafran War in the 1960s; served with Oxfam in Bangladesh, Sri Lanka and India; and served with UNICEF in Somalia, Uganda, Ethiopia, Sudan and, again, Bangladesh.

When Cole became UNICEF Regional Director for Eastern and Southern Africa in 1992, he and Marilyn took up residence in Nairobi.

And there they have remained. Cole, since his retirement from UNICEF in 2002, has been active as an adviser and consultant, and in institution building, in a number of programs—most notably Makerere University in Kampala and BRAC in Bangladesh. Marilyn has worked toward the rehabilitation of street children through the performing arts in a program she helped found.

Cole's work over the years has at times been routine. All too often, it has also involved high drama and extraordinarily complex negotiations. He worked on "both sides" following the civil war in Nigeria in the 1960s. He organized Oxfam's emergency and relief work among the great numbers displaced by flooding and facing starvation in the Rowmari area of Bangladesh in the early 1970s. With UNICEF's program in Uganda in shambles following Idi Amin's disastrous rule, he was called upon in the early 1980s to take over and return it to order, and did so, frequently under the most perilous of circumstances, always with great intelligence, courage and perseverance.

Cole organized a "corridor of peace" to the advantage of those tangled tragically, through no fault of their own, in the fighting between Museveni's gathering strengths in Uganda and the government, prior to Museveni's taking over. Later, in Sudan, he organized relief operations to assist those who were caught up, again through no fault of their own, in the war between the government in the north and rebels in the south.

Some would have found it impossible, or at least unworkable, to live with the many moves and intrigues and times of gunfire and coups and unrest that have come Cole and Marilyn's way. Not Marilyn. Not Marilyn, given her background. Not Marilyn with her character. Indeed, under just such conditions she has tended to thrive, drawing as she has simultaneously upon the faith and hope and love and resilience of people who are forced to live under such circumstances, at the same time helping them to make do.

Marilyn encouraged craftsmen and women in Uganda through the bleak and often torturous Obote and Okello years, visiting them in their villages, upon occasion supplying them with the materials they needed to pursue their crafts, and, with the help of war-ravaged coworkers, setting up her wonderfully successful Uganda Crafts outlet and coffee shop in Bat Valley, Kampala. Along parallel lines, she subsequently worked with craftsmen and women in Khartoum, helping to set up the crafts outlet Al Sudaniya there.

Marilyn is resolute, steady and efficient. She is also spontaneous and artistic. She loves colors and handicrafts, and over the years has brought to vibrant life the many homes in which she and Cole and, while they were home, their three children, Jessica, Sanjeeb and Jhorna, have lived.

Cole and Marilyn have never boxed themselves in with thoughts about what is and what is not possible, and never will. Rather, they carry on, like they have all along, a day at a time, always with the next dawn in mind.

The "big skies" of Africa, under which they have lived and worked and gone on safari countless times, to the most remote as well as more common places, suit them well.

The stories in this book come to us now in Marilyn's own storytelling style, rich with spontaneity and great bursts of color, the stories of a person who has never failed to embrace life without looking for the goodness of the Lord, whatever the circumstances.

Part One

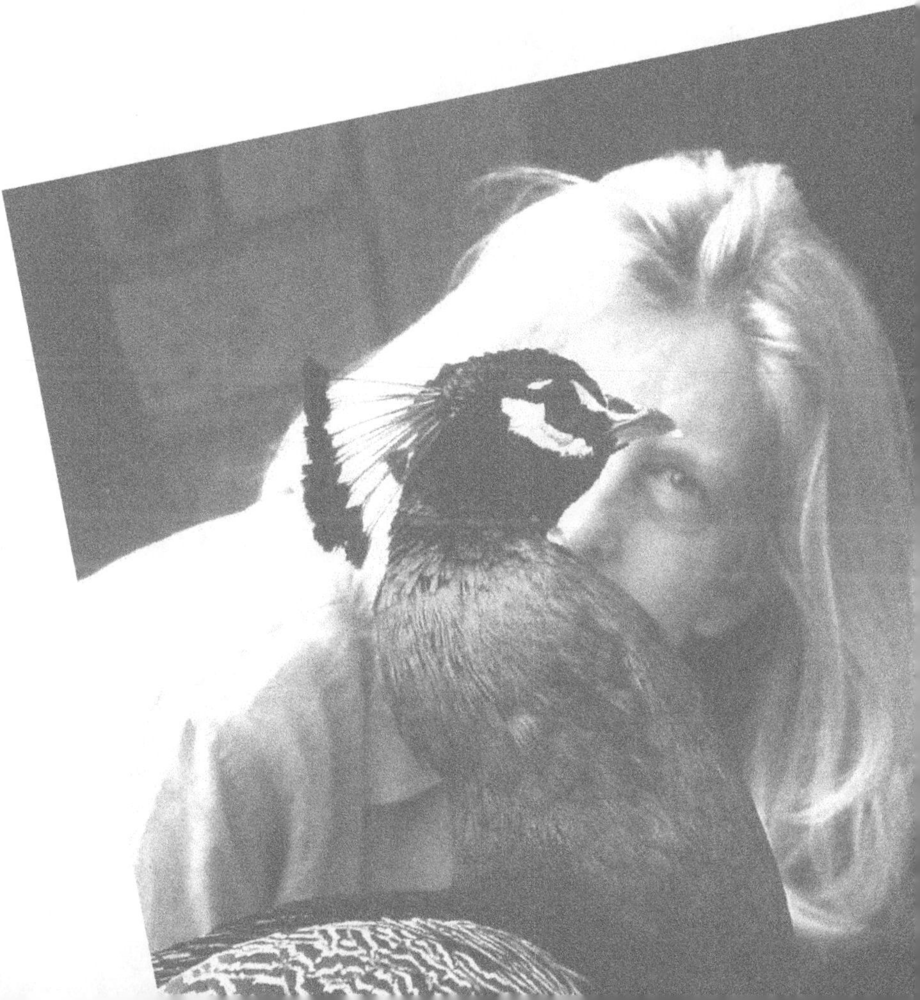

Prologue

As it was approaching dusk one day in August 1986, Kampala's streets were still wet from the late-afternoon storm that had swept over the city from Lake Victoria. Streams of muddy rainwater washed down the sides of the broken sidewalk into the numerous potholes. Near the entrance of the Nile Hotel, sentries took note of the blue United Nations flag flying from the wing of the car and directed Joseph to park under trees that dripped moisture.

A number of people were already assembled at the reception the government was hosting in recognition of Cole. It would be the last time we would see many of the friends we had made over the years. Above us, the windows of the Nile Hotel reflected back the late sun and I shuddered at the thought of what Penny had endured in the building that darkened in the fading light. For it had been on the third floor that she had been beaten and kicked in the mouth. Too traumatized to walk, she had crawled at the feet of her tormenters and begged for mercy.

Sobered by the thought, I pulled the shawl closer around my shoulders as a light breeze stirred the evening air. I greeted Elsie Kajubi and other women dressed in elegant Ugandan national attire tied with broad sashes. "Soon to leave us," Elsie said warmly as we embraced. Holding my hand, she reminded me not to forget old friends. "Not a chance," I replied, before I excused myself to join Francis Nnaggenda, who stood a little apart from the crowd. Francis's sculptures and art had captured Uganda's trauma like no other.

Professor Senteza Kajubi's resonant laugh sounded above the voices of those gathered as Daudi Talliwako, Chief of Protocol at the Ministry of Foreign Affairs, clapped his hands to call attention to what he had to say. Commencing his speech, he congratulated Cole for having been the longest serving Head of Mission in Uganda. "We will miss him," he said, reminding those gathered that Cole had successfully run the Corridor of Peace. V. Tarzie Vittachi had written about this in *Between the Guns, Children as a Zone of Peace,* stating that Cole had "formidable negotiating skills . . ." The first of the forty-six Corridor of Peace flights (the first time relief supplies had crossed the lines of civil war within a country with the consent of both sides), had taken place on 24 October 1984 following negotiations Cole had with Museveni and his NRA, who were fighting from the bush. Then, when the NRA came to power, Cole took up the issue of child soldiers. For, although Museveni felt the *kidogos* were looked after by their "adoptive" fathers, Cole felt war was the only thing the children knew and that it was time they attended school. Now, as Daudi Talliwako applauded my husband's accomplishments, I watched members of Museveni's government listen attentively.

Preoccupied with my thoughts, I felt Francis nudge my elbow when Senteza summoned me to the stage. "This lady," he said, embracing me, "and I worked together for the children with Child-to-child." He added that he hoped Uganda Crafts would continue to flourish. Cole was then asked to say a few words, after which the crowd started to disperse, for, although Museveni had been in power for six months, danger still stalked the night and they were anxious to reach their homes before the hours of darkness.

Cole and I lingered with a few friends to exchange recollections and well-wishes. Then, saying our final goodbyes, we walked slowly back to where Joseph had parked the car. A single fluorescent light lit the entrance to the Nile Hotel and, on sudden impulse, I asked Cole to give me a minute to walk through the rooms Penny had told me about.

"Better not now," he advised.

Only then did I notice the hotel lobby was full of soldiers.

It was not until the summer of 2007 that I had the chance to walk through what had been the Nile Hotel. It had taken eighteen months to renovate it into the Serena Hotel, which President Museveni presented as "a sure sign of recovery." Certainly the past had become memory, for the rooms that had been stained with blood, where Penny had collapsed at the sight of men beaten until they were unrecognizable, were now sumptuous suites overlooking terraces bordered by pools of water.

When Cole accepted the position with UNICEF in Uganda in 1980, the directive banned families from accompanying staff members to the country. The instruction was reasonable enough. Idi Amin's brutal rule between January 1971 and April of 1979 had led to economic decline, widespread social dislocation, terrible human rights abuses and the murder of between one hundred and three hundred thousand Ugandans. Instability continued after the Tanzanian army, assisted by Ugandan exiles and Libyan troops, liberated Uganda in 1979. Yusuf Lule, who became president after Amin fled, remained in office for only sixty-nine days. Lule's successor, Godfrey Binaisa, in an ongoing struggle over presidential powers in the country, was removed from the presidency in May 1980. A military commission chaired by Paulo Muwanga was in charge until the December 1980 elections, at which time Milton Obote was returned to his second term in office. The country suffered many changes that combined with the uncertainties of the reactions of different ethnic and other factions supporting the leaders in power to drive the nation further into chaos.

At first I objected when Cole heeded the advice of friends in Mogadishu and decided it would be best if the children and I remained in Nairobi, in neighboring Kenya, while he moved to Uganda.

"It's crazy," I said. "Why does the UN consider Kampala safe enough for you, but not for us?"

Cole watched Sanjeeb tease the girls. Unperturbed, Jhorna tossed her braids and put her arm around Jessica, whose curls turned golden in the sunlight.

"For the time being," Cole responded, "it's better this way. Kenya is stable, everything's available. The schooling is good for the kids . . ." Then, wrapping up his reasoning, he added, ". . . and I won't need to worry about your safety."

That very night, the grand old Norfolk Hotel in the center of Nairobi was bombed. It was New Year's Eve and we were planning to take the Oxfam Field Director out for dinner, when we heard that at least eighteen people had been killed and dozens injured in the blast that rocked the city. The Popular Front for the Liberation of Palestine claimed responsibility for the carnage. It was later concluded that the PFLP had attacked the hotel not only because it was owned by the Block family, who were Jewish, but because Kenya had allowed Israeli commando units to refuel in the country four years earlier on their mission to Entebbe, Uganda, to rescue mostly Israeli passengers from a plane hijacked by Palestinian terrorists.

We had only a few days to find a cottage in Nairobi before Cole proceeded to Kampala. Making do with minimal furnishings, the children and I settled into routines in our comfortable house with a lovely garden on David Osieli Road. We were familiar with Kenya, because, when we had lived in the Sudan and Ethiopia, the country had been our chosen destination for holidays, with camping safaris to wonderfully wild parks, the coast, mountainous areas and along rushing muddy rivers. A friend and I took turns car-pooling the children to their classes at Kestrel Manor, which was directed by a British woman, Mrs. Byrne. There, dedicated teachers introduced the children to facets of Kenyan life, including the various tribes, with trips to outlying areas. Two or three times a week, I ferried the kids to parties or they had their friends over to play with the two longhaired kittens we had brought home from the Kenya Society for the Protection and Care of Animals. We were a stone's throw away from Westlands Shopping Center, with well-stocked commodity shops where we could purchase every type of grain, fruit and vegetable. Most days were cool and sunny, and I delighted in the flowers that grew in our garden and always kept a large bouquet on the table in the kitchen.

But we missed Cole. Thrown onto my own resources, I searched for footing when responsibility fell heavily on my shoulders. And although the clean high-altitude air and grass plains of Kenya softened my loneliness, within six months we decided that we could no longer live apart as a family. Despite the violence and fear that featured so much in the lives of the Ugandans, and Obote's rule slipping further and further into what would later become known as the massive abuse of human rights, Cole came to Nairobi to help us travel to Kampala.

Although I begin this book with experiences in numerous other countries, much of what I write is about our time as a family in Uganda in the years between 1981 and mid-1986 and stories of our friends there. They were turbulent times, with many of our days edged by incidents like the one related above. Insecurity was often a frightening reality, with disappearances, tortures and killings experienced all too often by persons we learned to know well. Curfews were imposed. Sometimes we learned too late what it would have been good to know before we ventured out.

Roadblocks manned by ill-fed, ill-equipped and often ill-mannered troops dotted the thirty-five-kilometer road between Entebbe and Kampala (we once counted eight roadblocks), each with its own uncertainties. Very few commodities were available in the few dry goods stores still open. Hotels and restaurants were no longer functioning. Garbage collection was nonexistent, and telephone and medical services were not reliable. We had a backup generator, as the supply of electricity was sporadic at best. Our water supply was delivered by truck from Lake Victoria under our own arrangement and stored in a large metal tank near the back of our house.

This story is partly about living under such conditions. But it is also about Uganda's wonderful people and their strength, resourcefulness and resilience to survive their personal hardships under the most terrible of circumstances in the beautiful country Winston Churchill referred to as "the pearl of Africa," now ravaged by war.

Chapter I
Indian Beginnings

I KNEW THE HOT dry plains of Andhra Pradesh when I was very young, where seasonal dry riverbeds meandered near soft green paddy fields planted at the base of granite hills. Under the shade of neem trees, oxen pulled carts laden with people and rice straw.

In June 1951, Mother accompanied me on a crowded bus up the misty Palani Hills to boarding school. Eight years of age, I had been coddled by my beloved family, and the prospect of leaving my parents for the first time filled me with apprehension, but also with some anticipation. Hanging back on my mother's arm, I blinked in the sunlight that splashed on the coir matting running the length of the hallway of the dormitory that would be my home. People I had not met before hugged my shoulders, and excitement prickled my arms when chattering girls helped me fold my clothing into dresser drawers in the room we would be sharing. Tucking the sheets and blankets under the mattress, we spread the familiar green quilt appliquéd with white elephants over my bed. But just as I began to feel appreciation for my new friends and surroundings, fear clutched my heart and I clung to Mother and what I knew. Meeting my tear-filled eyes with her moist ones, Mother gave me one last hug and pulled the heavy sweater about her shoulders in a determined manner before she left me surrounded with God's blessings and what she prayed would be good care. Heartbroken, that night I sobbed into my pillow long after the others had gone to sleep.

But my tears soon dried, because the dormitory was filled with children who became my friends. Terraced flowerbeds shaded by a spreading jacaranda tree descended to an open field in front of the dormitory, which was situated on the edge of a hill overlooking the lake and the picturesque boathouse. Throwing our books and bags in our rooms after school, we played house in the gnarled old roots and limbs of an ancient tree or on the platform built among the branches. And if we could capture shiny black beetles, we built tiny structures for them on the moss carpet of the forest floor until called to wash up before dinner. Following prayers and a story, the housemother bid us good night. Although it was forbidden, we whispered long after lights out, or read Nancy Drew and Hardy Boys mysteries by the glow of torches under our bedcovers.

The school we attended was located on the brow of a hill across the bund that controlled the water from the still gray lake. Near the stone wall on our one-kilometer walk to school, the *halwa* man always beckoned my best friend and me to come and see the sweets he had made. Opening the lid of his glass case to tempt us with peanut brittle, chewy *halwa* sprinkled with nuts and green mints covered with chocolate, the old man, whose tooth hung limply over a cracked lip, coaxed us: "Missimma, you buy one . . . or two?"

Nibbling a portion, I tucked the rest of the sweet into a pocket, where it was frequently forgotten . . . until the day the *dhobi,* washer man, complained to our dorm mother of my sticky habit.

"Twelve, seven, two," I counted to myself, smudging another day off the November calendar . . . until it was time for the dormitory *ayahs* to braid my straw-colored hair into pigtails and our "party" boarded the local bus to wind down the *ghat* road. At the base of the hills, I welcomed the smells and sights of familiar village scenes. Goats stripped the leaves off brush alongside the narrow road. Honking to clear the way of bullock carts, the bus driver parked at Kodai Road Station and called coolies to help carry our luggage to the train that halted long enough for us to hurry toward our reserved compartment. "Dubsing" my bunk for the night, I spread my bedroll on a lower one, from which I could watch vendors hawk their wares and see the city of Madras when the train slowly pulled into Central Station the next morning.

Following another night on the train, I braced back the heavy door of the compartment and leaned out into the dry air to watch the rocky hills and scattered bush of Andhra go by. Squinting in the white rays of the noonday sun, I caught sight of two dearly beloved persons standing together on the platform of the Mahbubnagar Station:

Mother with her hair pulled back in a bun and Dad with one shoulder lower than the other, both wreathed in smiles—hands aloft, waving.

Impatient for the final screech of brakes as the train slackened its speed, I jumped down to the platform, wild with happiness and full of talk; ready to be home and feel the earth and cool stone of slab floors under my bare feet.

Early sun streamed through the screens of the front verandah where, surrounded by pots of plants, breakfast was served. Following morning prayers, I accompanied Dad or Mother to the school, church, dispensary or dormitories they administered along with countless other unexpected tasks; or tucked my knees under me on a chair near my father in his office. He always made time for me when he was there. When he wasn't, I swiveled slowly in his worn chair—where the smell of Quink ink and typewriter ribbon were part of a safe and familiar world—round to face the framed pictures on the wall: a smiling Dwight D. Eisenhower next to a portrait of Mahatma Gandhi bent over a stick—past the filing cabinets, and back around.

In the mornings, I would often follow the cement rim of the roof, accompanied by at least one of our cats, until I reached the highest pinnacle of the bungalow. Closer to the clouds that patterned the vivid blue sky, I watched the *gubilum*s scatter in confusion when schoolboys, who were paid *anna*s for the task, banged empty tins to frighten them away.

Barefoot most of the day, my feet became hardened and brown like those of Pauquim—our cook's youngest daughter, who ran with me through the stalks of grain left to dry in the field. Other mornings, I followed my twin brothers to the big well, where only after an empty twenty-kilo oil tin was strapped to my back to act as a float was I allowed to venture down the steep steps to the stone ledge below. Straining to keep my head above water, I dog-paddled much like the large frogs I tried to avoid. Shivering with cold and eager anticipation, I watched David and Paul crawl up the thick trunk of the banyan. Up and up they climbed, to a branch that extended overhead to dive with narrow flashes into what seemed the dark recesses of bottomless depths. Holding my breath until they resurfaced, I crawled up the rock steps to the top of the well, where Balayya, his *puncha* tucked between his legs, prodded the great white team of oxen with a switch. Twisting their tails to coax them backwards, he repeated this procedure to lower the leather bucket into the well again and again. Water splashed over the ropes the oxen pulled over the wheel before it was released into the earthen channel to irrigate the fields.

The ayahs who worked at the girls' dormitory welcomed me into their smoky kitchen. Squatting near the fire, I persuaded the cooks to spoon *rasam* out of the oversized *dekshi* into a tin plate. Favored as the *dorasanni*'s little daughter, I mixed the fiery hot sauce together with the *sankati* they had cooked and slurped up the peppery sorghum. Licking my fingers clean, I wandered back to the bungalow and climbed up the granite stairs to the washroom, where, sitting cross-legged on the cement floor, I poured *chembu* after *chembu* of cool water over my head from the waist-high *ouzu* in which the water was stored. Then, curling up on a cushion in the shade of the verandah, I lost myself in Louisa M. Alcott's series or Robert Hardy's *Seven Last Days* until clatter from the kitchen below indicated preparation for meals.

In the kitchen, Mariamma allowed me to pound spices in the rock mortar. Trickling water into the smooth hole with my left hand, I pounded the pestle with my right until, impatient with my play, she nudged me aside. My Telugu friends and I were often scolded for crushing neem berries and leaves in the mortar. But Daniel, who had been trained by Mother to prepare our meals, allowed me to flatten moist patties of coriander, cumin, ginger, cloves, cinnamon, asafetida, garlic, mustard seed, turmeric and chilies on the aluminum lid after they had been ground. After combining the fragrant spices, he let me stir them into the finely sliced onions he then made into the most piquant of curries on the wood-burning stove.

At the close of day, I crawled onto my wide spring bed and pulled the covers over my head, until Mother came up the stairs and chased away the shadows that swept with whispers through the screens of the open doors leading to the verandah. The great room I slept in was similar to those in the bungalow the *Raja*, king, had assisted foreign missionaries build some distance away in the village of Wanaparty. There the walls of the interior of the bungalow had been coated with lime mixed with the whites of thousands of eggs—something that troubled me as a child. *What had happened to the egg yolks?* I wondered as I ran my hands over the smooth shiny surface of the walls, which glowed ivory with age.

On special occasions, school workers loaded ox *bundi*s with cooking vessels and ingredients necessary to feed five hundred mission-school students, staff, their children and church workers. While the oxen plodded slowly along dusty roads, other children and I jumped from the carts to run alongside and pluck ripe cactus fruit. Reaching our destination—the ancient spreading banyan, which resembled a low hill from a distance—those of us young enough to avoid responsibilities swung like the resident monkeys on the twisted hanging roots of the revered tree, until called to sit in a large circle under the shade of the *marichettu* to enjoy savory rice and mouth-watering chicken curry served to us on leaves stitched together with split reeds.

If my parents found a day of free time in their busy schedules, Daniel packed containers with curry and chutneys, together with a *dekshi* of rice, utensils and plates, to take to the Jubilee Hills. The musky smell of panthers scented the warm air as we drove down the dirt track snaking near a gently flowing stream. Spreading mats under the forest canopy, Father reclined on the *sapa* next to Mother, who rested her head on a bent elbow, and they would doze—part of the dappled sienna shadows of an Andhra afternoon. And so the sunset of that day turned into the sunrise of the next and the years became a rich background of memories.

Return

At the age of fifteen, I traveled to America with my parents. When they returned to India, where their duties included teaching and administering to seminary students, I stayed behind to further my studies. Late in December of 1963, my mother buried my sixty-three-year-old father, and her partner for thirty-seven, not far from Ramapatnam on the eastern coast of India. Local fishermen had brought his body back to the shore through waves that swept him into a turbulent Bay of Bengal. The casuarina trees stood darkly against the sky as a human wall of villagers surrounded my mother and eldest sister Esther with support the evening they labored over Father's form—accepting death only after hours of resuscitation and massage failed to revive the "beloved and respected *Doragarru.*"

The drooping silhouette of branches framed the sea whose restless waves echoed the ebb and flow of life.

Tenderly washing and dressing his body before hundreds of mourners came to pay their final respects, Mother laid Father to rest in a graveyard under swaying coconut palms and *chinta* trees.

Following my father's death, my brave and beautiful mother chose to continue her work in India. Arranging for me to leave the University of Kansas to spend time with her on the seminary grounds in Ramapatnam, she found work for me to do in the office there. Well before sunrise, Mother had her devotions, and I would hear her make her way down the stairs to her dispensary in order to attend to a *campu,* birth, or see someone in distress. When sea breezes relieved the heat of the day, we walked to the fishing village to watch the waves pound the sand and talk with those who knew her well.

Just a train ride away from the Ramapatnam seminary, my eldest sister Esther (who had taught me songs late into the night when I was a child) tended to the sick at another mission station. The hospital where she worked was depressingly neglected when I visited her in Hanamakonda—the hill of the monkey god, not far from the temple ruins of the Kakatiyas.

Huge old trees rustled seeds bleached by the relentless sun that extended heat through the long drought of that season. Oblivious to the dried grass that pricked their bare feet, children ran over the parched earth, past water buffaloes listlessly switching their tails at the bothersome flies. In the midday heat, Unkamma, who worked for Esther, cradled a little one in her arms and crooned "*Adaya, daya, dey*" in a low throaty voice that echoed down the dark high-ceilinged rooms.

One day, I heard Esther shut the screen door of the two-story bungalow she lived in. Closing my book, I followed her to the nearby clinic, where the matron straightened the white cap pinned to her black hair and showed us into the examining room. Seated on a stool, a small girl gazed trustingly at the doctor, who peered closely at what he knew to be the cause of the child's discomfort. Outside the door, patients waited patiently to see the doctor, who took off his white coat and announced he would return to his duties later that day.

That night cries woke me from the adjacent bedroom. Pulling aside the curtain, I watched Esther drip warm milk into a tiny infant's mouth with a syringe. Carefully wrapping the child in a blanket, she turned on the light bulb just above his cot to keep his little body warm. *Just like handling a baby chicken*, I thought, as Esther repeated this procedure with the little one hour after hour—night and day—no matter what other duties she had to perform. Devoting herself to other abandoned babies brought to her by villagers, in time my remarkable sister found homes for dozens of children she nurtured—three of whom she adopted as her own.

The last of the white medical missionaries to work in Hanamakonda, Esther frequently left the unhappy hospital and its staff and drove her van and helpers to the surrounding villages, where she held clinics for those who, as my brother Paul writes in *Sepia Prints*, "used potions and mantras to protect themselves from evil and sickness . . . [and] amulets, which assured support more effective than the supports they could assure for themselves." It was to these people, who knew the covetous eye of another could be warded off with *disti bomas,* protection-giving images, that Esther offered her own potions and powders to aid in healing the villager's afflictions.

One sultry afternoon, we loaded the van and bumped over village roads to join a group of *Banjara* women seated under the shade of a *kodipilla* tree. Dressed in colorful skirts, with bracelets up their arms, the women were easy with their talk and laughter and showed us the items they had been sewing. Examining the tiny stitches with interest, Esther invited the *Lombardis* to come to her bungalow with their little ones to hear the word of God and learn how they could earn their own wages. Addressing them in fluent Telugu, she encouraged them to bring others of their tribe to embroider their customary beads, mirror fragments and cowrie shells onto brightly colored cotton cloth. In time, the pieces the women completed were made into beautiful pillow and bed covers that were sold for profit.

Ominous monsoon clouds swept black curtains of rain toward the bungalow Mother lived in. Great drops from the banyan trees splashed onto my cotton sari and spattered dust into rivulets of mud that coursed along the path I walked. Pulling the *palla* of the sari closer about my shoulders, I pushed open the door of the Ramapatnam seminary library office and sat down at my small desk to complete a letter to Rev. Solomon, who pastored the church nearby. Then, slipping a fresh piece of paper into the Remington, I typed the address of the American Peace Corps to make inquiries concerning volunteer work in India . . . for I wanted nothing more than to remain in the country of my birth.

Chapter II
The Art of Peace

EARLY IN 1965, Mother and my sister Esther accompanied me by train to Bombay (Mumbai), where I boarded a ship to sail back to the United States to continue my studies. Two months later, I received a letter of acceptance to train for the Peace Corps in India in the state of Texas. The Peace Corps trainers informed us that the subcontinent would be overrun with poverty, illiteracy, land fragmentation, starvation and political chaos by the end of the century. Confident in their predictions, the trainers dismissed Asian goals for development, food sufficiency and industrial competence as a mere wish list. But I thought otherwise . . . and, fortunately, they were wrong.

Following the completion of courses and the selection process, we flew in a chartered Pan Am plane to Palam Airport in May 1965. It felt good to feel the warm embrace of Mother India again—a country rich with ancient culture, peoples, philosophies, religions and languages.

Heavy wet heat wrapped about me, although it was still gray dawn. After completing our immigration procedures, we boarded a chartered bus and the driver eased onto one of old Delhi's crowded thoroughfares. Pushing back the tinted window, I watched throngs of shoppers carry their merchandise past walls painted with BE INDIAN BUY INDIAN: products purchased by a rapidly increasing and prosperous middle class.

Along the sides of streets, huge billboards advertising Bollywood's latest films shaded makeshift dwellings. Overcome by the exhaust fumes that choked the air, honking horns and din of traffic, I closed the tinted window as the driver drove the shuddering bus towards the relief of stately old trees that shaded the Maiden Hotel. Porters dressed in white uniforms greeted us at the hotel's arched entrance and helped us carry our luggage up stairs bordered with pots of yellow marigolds.

Wearily sagging into the wicker chairs grouped on the wide verandah, we waited to be allocated the rooms we would be sharing.

That afternoon, ceiling fans rotated hot air around the large room as Peace Corps staff read out the locality we had been assigned to. Although pleased to be going to Bangalore, a city in Karnataka with a favorable climate that attracted one of the fastest growing populations in India, it was against my wishes that I was to perform secretarial duties for the director of South India.

"I know nothing about office procedures," I pleaded with the personnel in Delhi. "And anyway, it is rural India where projects translate into the sort of development issues I would like to be a part of." But alas, those in charge were convinced my childhood familiarity with the subcontinent would cater to the needs of the volunteers. My only hope then was to change the mind of Kellogg Smith, who administered the office in the southern city. Soon after, I was given a train ticket to travel to Bangalore.

The elderly woman easily crossed her legs into a lotus position on the seat of the train's second class compartment we were sharing. Smoothing back her thinning gray hair, she reached for the *tiffin* carrier under the seat and pried open the stainless steel containers, offering *chapattis* and spicy vegetables to her husband and then to me. Belching appreciatively at the end of his meal, the old man gulped water from a mug before he rinsed his mouth and spat the remainder through the iron-barred window. Then, spreading a thin cotton sheet on the narrow bunk, the old woman lay down beside her husband and the two of them fell asleep while I reflected on the easy acceptance and generosity of my traveling companions and watched the sun set over dusty villages.

It was nearing 2 a.m. two days later when the Grand Trunk Express halted at Bangalore's Cantonment Station. Tired and disheveled, I stepped around shrouded people stretched on the platform and made my way to the ladies' waiting room. Every inch of space was taken up with women sleeping with their children tucked near them on the crowded floor. Too tired to care, I crawled atop a large table and fell deeply asleep until an *ayah* switched on the overhead light. In need of a wash, I made my way to the latrine that reeked of camphor and stale urine and held my hands under a leaking faucet. Splashing water on my face, I dried with the musty-smelling towel the attendant in a starched white *sari* handed me before extending her hand for baksheesh.

"I hope you are enjoying India?" she inquired with a friendly smile.

Nodding, I pulled my suitcase out the door and pushed my way through the crush of early morning travelers.

"*Coolie, ma?*" the porter inquired, before hoisting my suitcase to his head. Grabbing my hand luggage, I hurried behind him to one of the motorized rickshaws parked near the curb. Impatient with the coolie, who placed my luggage beside me on the seat, the driver revved his motor before he drove us through Mysore's sprawling capital, past the Kalakshetra auditorium and on towards Richmond Town, where trees dropped Ghol Mohar blossoms on the pavement below.

I had known Bangalore before. However, "You can always rediscover an old path and wander over it," as Beryl Markham writes in *West with the Night*, "but the best you can do then is to say, ah yes, I know this turning! Or remind yourself that, while you remember that unforgettable valley, the valley no longer remembers you."

In time, I found a place to live to the side of an old Anglo-Indian woman's house, a bicycle ride away from the Peace Corps office on 21 Alfred Street. Caring and devoted, little Mary found work with me. One evening, reproachful that I had left half a *chapatti* on my plate, she left me with safe blessings and a reminder to slide the door bolt into place—instructions I made light of until, glancing up, I saw a dark face peering at me from the transom. *How long had the man been watching me bathe?*

Heavy rain fell the next morning when, unmindful of the candle that dripped, Mary helped me push hot wax into the crude holes the intruder had drilled round the lock of the door. Unusually sober, she patted my arm and called out that she would be spending the nights with me as I left the house to cycle away in the comfort of new sun.

But the disturbances continued, and I came to dread the hours of darkness and what I thought were the prowler's intentions. Hearing of my dilemma, Albert Franklin, the American Consul in Madras, instructed the Peace Corps office to have police stationed at my residence. Thereafter, boys with oil-slicked hair called out, "Hi, honey!" with comments I largely ignored when cycling to work to respond to the expectations of Kellogg Smith who, having worked as an editor of the *San Francisco Chronicle*, was meticulous about every task.

Fed up with long hours and my deficient secretarial skills, one day I pushed back my typewriter and expressed my wish for change. Reacting to the tone of my voice, Kellogg slumped into the chair next to mine and listened before he pointed to the bulletin board above my desk and the poem his wife, Peggy, had written.

There was a young girl in Bangalore,
Who promised two years to the Corps.
After a month and a day, of hard labor for K,
Said, "Good grief, it's not peace, this is war!"

"I know I'm tough but I have no intention of letting you go," Kellogg said with a smile. Placing his spectacles on the table, he rubbed his eyes with both hands and told me the office had work to do and that I had an important role to play.

Convinced he had not taken my request seriously, I thought to share my feelings with his wife. Normally energetic and tanned, she appeared drawn and pale a few mornings later when she leaned her bicycle against the wall of the Peace Corps office. Slowly climbing the stairs, she swayed unsteadily towards me. Stooping to pick up the bouquet of asters she dropped to the floor, I told her Kellogg was due to arrive from Delhi later in the week before I inquired as to whether she was unwell.

"I'm all right," she replied faintly.

But I thought otherwise and called out, "Take care," as she left me to ride crookedly down the road beneath trees with flowering branches.

The next morning the voice at my door urged me to "Come quickly." Hastily pulling on my clothes, I pedaled through Richmond Town's deserted streets to find Peggy's visibly distraught cook wringing his hands near the entrance of the Smith's residence. Propping my bicycle against a pillar of the verandah, I reluctantly followed him through the living room, past end tables with half-finished bowls of cashew nuts and tight bouquets of purple and pink asters, fearful of what I had already suspected. It was dark in the bedroom. The curtains had not been drawn. Taut with concern, I walked to the bed and leaned over to touch my friend, only to recoil. Peggy was stiff with death. Pulling the sheet over her face, I stumbled out to the light of the garden, followed by the distressed cook, and sobbed under the tree.

Unsure what to do, I slowly cycled to the Peace Corps office, where I found Mr. Tapp, the administrative assistant, pacing the floor. Exchanging sober greetings and what we knew, he picked up the phone to call Kellogg in Delhi to inform him of the tragic news. Clearly shaken, Kellogg's response was to dispose of Peggy's belongings and have her body cremated.

Peggy's body, wrapped in a sheet, was placed on a pyre of neatly stacked wood. Crackling flames licked the night, and cast flickering light on those of us gathered near. I swallowed the emptiness I felt, and tears washed the heat from my eyes as smoke lifted and the body charred into ashes. No words or prayers were spoken, and I sobered to the reality of that heartless occasion, followed by days with learned directives that catered to the needs of the volunteers.

One of those volunteers quickened my heartbeat.

"Despite challenging others, Dodge is usually correct," I heard Kellogg mutter to himself one day as he scanned the comments he had written. Leaning back in my chair, I smiled to myself and reflected on how the atmosphere became charged when Cole entered the room. Blessed with good looks and a rational mind, he pursued his commitments with a great deal of energy and deliberation.

Not long after, I welcomed Cole's invitation to have fresh lime and soda at Koshy's, where, seated opposite me in a booth, he focused his blue eyes on mine and proposed marriage. Taken aback, I turned my head to conceal a smile, and suggested we get to know one another first.

Then I slipped my hand into Cole's, and we walked down Brigade Road, which was bright with neon lights, to a Brahmin restaurant we knew served excellent food. Generous with their welcome and portions, waiters served us hot *puris* we dipped into steaming *sambar*, dhal soup, and vegetables, followed by as much of the same as we wished. Spooning *ghee* over the steaming rice they heaped on our plates, men served more *rasam*, fresh pickles and curries, until, satiated, we covered our plates with our hands, indicating we had had enough of the tasty food. Delighted that we chewed the betel nut offered to us at the end of the meal, the man at the cash register wished us well as Cole summoned an auto-rickshaw to take us back to Richmond Town.

Ducking his head under the rickshaw's canvas covering, Cole sat near me on the weathered seat. "Just last week," he began, as we bounced down the street, "I was riding behind my friend Krishnamurthy on the back of his motorcycle when he hit a bump and I was thrown off." Chuckling, he said he hadn't been hurt but had landed on wet cow dung that caused him to slide down the hot tarmac. "Despite the accident, we had a good time in Mysore City. You know, Peace Corps for me was a godsend," he admitted. "Something that saved me from a disastrous launch into the redneck world of Spokane, Washington."

Confident, and a gifted creative thinker, Cole left his poor rural home in the pine-covered stony hills of eastern Washington at the age of fourteen and worked as a farmhand. One day, following a fistfight he had with the student body president in the high school he attended, he was called to the principal's office. Expecting a reprimand, instead he was told that the president of the United States had been assassinated and would he, as the senior class president, address the student body?

Cole told his audience that he planned to follow what Kennedy had instructed and ask not what the country could do for him but what he could do for the country. Shortly thereafter, he sent his application in to the Peace Corps.

But before Cole received his letter of acceptance, at the age of eighteen, he followed the winds of chance and change and set out from rural Washington to find his fortune or a job: "whatever came first." Tucking two hundred dollars into his pocket, he purchased a one-way bus ticket to the bright lights of Los Angeles and rented a room from a kindly black woman in the central part of the city.

One sundown, after walking from one temporary employment agency to another, he wandered into the seedier streets of LA. A group of Hispanic toughs flashed their knives and cleaned out what he had in his pockets. Scared, roughed-up and out of breath, Cole ducked into the entrance of the hostel and asked the sympathetic landlady to refund enough of his money that he could buy a return ticket to a place where he understood the community. With enough coins left to feed the machine for a phone call to his mother, he boarded a bus back to Spokane, where he found a letter inviting him to train for the Peace Corps in India.

In the months that followed, Cole taught me how to be generous with my time and talents. Before I got to know him, I had not been stingy about caring, but our bond widened my heart and taught me to pursue things I felt keenly about. As an artist, my romantic sentiment helped to soften Cole's edges, and we found that being together enriched both our lives. Lives that we agreed to join in marriage in December of 1966 in the hills of Kodaikanal.

Our journey up the hills began in earnest at Kodai Road Station, when the train screeched to a halt at the base of the Palani Hills in Tamil Nadu. Noise and confusion reigned, as men with shaved heads and vermilion on their foreheads pushed their bags into the compartment, determined to find seats. Gathering our belongings, Cole jumped to the platform and grabbed the suitcases from me until coolies hoisted our luggage to their heads and led the way to a taxi draped with dried marigolds. The honeymooning couple seated at the back of the van shyly moved over to make room for Cole and me to position ourselves on the worn seat. It took time for the driver to adjust his rearview mirror and begin the honking ascent to Kodaikanal, six thousand feet above the plains of South India.

Mist lifted over the foliage to the sides of the road that wound around hairpin bends past wooded ravines, where sometimes monkeys swung from the trees. Now and then, the driver halted to cool the van's engine, and we purchased green coconuts from the vendors seated by the sides of the road. Sipping the sweet coconut water through straws, we marveled at the view until we were bothered by the tourists from nearby towns, who leaned from the windows of buses to toss litter across the scenic slopes. We were later amused to hear one disgusted observer say, "They piss on the hill, disturb the peace with loud cinema music and trash sites near spectacular views." Still, Kodai Conservation Council struggled to preserve the hills' delicate ecological balance, which boasted hundreds of species of orchids and rare plants. KCC posted signs on the *ghat* road exhorting people to "Save our trees" and "Don't let Kodai go up in smoke," and erected steel fences in an attempt to protect the receding *shola*s and indigenous forest.

White clouds covered Perumal's lofty hilltop as we rounded the last bend to Kodaikanal, the hill station where I had attended boarding school for ten years. Situated on a rise near the lake, the school opened its doors in 1901, with eight classes and a home for thirteen children in the Highclere Hotel until they could purchase the property. In time, as Nora Mitchell wrote in *The Indian Hill Station*, British and American civil servants and missionaries who spent hill seasons with their school children started a settlement there.

Now a popular tourist destination, Kodaikanal was built up with three-star hotels that violated the town's building bylaws. Pedestrians, a few cows and emaciated pariah dogs crossed the intersection known as Seven Corners. In time, the driver found a place to park and climbed onto the window frame to help Cole pull our suitcases from the top of the van. Nearby, I was sure, was the same cobbler, still sitting cross-legged, bent over an assortment of rubber and leather bits.

"*Repairii, ma*?" he inquired, studying my shoes with weak eyes behind thick spectacles. Not remembering.

Crouching, I reminded him of when I had complained to him in my younger days of shoes I thought were not well-polished.

"*Aamaa*," he responded hesitantly. Peering closer, he gave me a wrinkled smile that showed some sort of recognition.

"Have you work?" he asked Cole, who told the old man he would bring him his sandals to mend before he led the way toward the lake road. Perumal (named after the mountain), who had done odd jobs for my mother from time to time, slouched down the road beside me, whining of faded recollections and ". . . when you and church people were generous." Shivering convincingly, he pulled the worn blanket around his shoulders and told me of his sudden gripping fever, to which I responded by pressing a few rupees into the old man's hand. Perumal straightened his posture and smiled at Pushpa, who had *paan* pushed to one side of her mouth. Pinching my cheek, she showed me the tiny monkey she kept under the fold of her *sari*, then held out her hand for money before she left me for more lucrative prospects. Some days later, when I reproached her for not wearing the *sari* I had given her, she gave me a tender look that told me I should understand her profession did not allow for a changed appearance.

But Pushpa wore her new blue sari the afternoon of 30 December when my brother David escorted me down the aisle to join Cole in matrimony. Seated at the back of the Highclere chapel, next to Perumal, Pushpa beamed at me as the sun slanted rays on the proceedings. On the window ledges were large bouquets of deep red rhododendrons and white lilies, which family members picked during our excursion to the beautiful surrounding hills. Hopeful, and with a corsage pinned to her white suit, Mother presented my hand to Cole. In attendance were my three brothers and their wives and families, my sister Esther, a few colleagues and some Tamil friends. Paul then ministered to us from the beautiful thirteenth chapter of Corinthians, after which Cole and I walked away from the chapel arm in arm. Garlands were placed around our necks and flower petals strewn at our feet before we signed the register as Mr. and Mrs. Dodge.

The next morning, thick fog covered the view of Periyakulam and the plains below. Cole and I strolled down Coaker's Walk, a path that wandered just below Van Allen Hospital on the brow of the hill, where Mother had delivered me on 18 August, 1943. Farther up the hills, Cole and I hiked near granite pillars that rose vertically into clouds, past forested gorges and clear streams, where occasionally buffalo were spotted on an open *shola*. A week later, we boarded a bus to leave the highland air and travel to where we would work together in Mysore. We were entering a culture steeped in tradition, which boasted a rich history of dynasties known for fine silk, sandalwood, palaces and intricate temple architecture. And we were to meet the fascinating people of this culture.

"*Yelli*?" the man inquired, showing betel-stained teeth in a wide grin. Cole delighted him by replying in reasonably fluent Kannada (one of the major Dravidian languages of India, spoken mainly in the state of Karnataka). Clad in a *lungi* wrapped around and between his legs, the man encouraged Cole to say more as he helped us carry our luggage toward the weary horse that drooped its head beside the *tonga* it pulled. It troubled me that we were loading two heavy suitcases and ourselves behind the driver. Other wasted animals stood listlessly by the carriages they drew, near half-starved dogs searching for scraps in the rubbish heap thrown to the side of the Mandya bus depot. Brooding over animals left to their fate, I climbed onto the horse carriage beside Cole and asked the driver not to use his whip. The *tonga wallah* smiled and obliged as we swayed and jolted down the seven-kilometer dirt lane in and out of traditional villages, through sugar cane fields and rice paddies. I was back in rural India where I wanted to be.

Our first married year, Cole and I planted rows of corn and tomatoes near the concrete house provided by the Gramsevak Training Centre (GTC), where we made our home. We also purchased a buffalo that calved and provided rich milk. Although a monkey amused us for a while, we found it another home when it harassed the chickens we penned to supply us with fresh eggs. Most mornings, Cole bounced over back roads in a jeep. But occasionally, he and the Block Development Officer hopped on Cole's Enfield motorcycle to convince villagers identified under the Applied Nutrition Program about the management principles he had been taught in training and to advise them how to grow kitchen school gardens, organize midday meal schemes and inoculate their pullets on the exact day prescribed in the books.

Fortunate to have been placed as volunteers near one of the largest and best-managed farmer's cooperatives in the area, Cole and I found that, ultimately, local institutions did the work for the profits they gained. Ideal for the ingredients of poultry feed, the byproduct from the Mandya rice mill and peanut-crushing unit, mixed with dried fishmeal, was sold to farmers throughout the state of Mysore. One day, Cole strapped a sack of feed behind me to deliver to a local farmer and drove the motorcycle through villages where laughing children took shortcuts to reach the lanes we traveled. Blocking the road ahead, men prodded livestock decorated with limes pierced on their horns and garlands draped about their necks. Anticipating what followed, we stopped to watch men leap like satyrs through flaming rice straw as they drove their animals wild with fright through the smoke and fire that licked at heels and hooves to celebrate the harvest festival.

Fascinated with our response to the colorful festivities known as *pongol,* children accompanied us to the center of the village, where Cole delivered the grain to one of the poorest of the poor caught up in the cultural net of tradition and survival routines, who owned a small backyard unit consisting of a dozen or so cackling hens. Better-off families kept one or two hundred hens, while an ex-army officer and a retired civil servant set up large poultry farms with up to a thousand chickens. Collecting the eggs weekly, Cole and his colleagues loaded the baskets on top of a local bus and took them to the vibrant wholesale market in Mysore City, where the price they received was not enough, but the eggs were transferred into purchaser's boxes and resold.

While Cole promoted the practice of poultry, I held the hands of small boys and girls with sweet-smelling flowers braided in their hair and walked to the *balwadi.* Eager to play games, sing songs and teach me rhymes, the children at the preschool crossed their legs on mats until midday, when they were served a meal of corn, soybean and milk powder mixed into gruel. And when the children were not feeling well, I accompanied them to their mothers, who showed me the herbs and plants they boiled to treat their little ones' disorders. But the potion a friend gave me to relieve my congested sinuses offered me no relief, so Cole suggested we drive to the city of Mysore for treatment.

The sign above the door read *Ear, Nose and Eye Specialist,* followed by the various degrees the doctor had earned or failed. Entering the hall, which was lit by a flickering fluorescent light, we climbed the stairs to where a kindly old man greeted us with hands pressed together. Inviting us into his modest treatment room, he offered me a seat on a high stool and peered at me over eyeglasses that rested at the end of his nose. Listening attentively to my complaint, he opened his desk drawer and rummaged through paraphernalia to find a long red rubber tube. Wiping the tube with a well-used towel, he inserted one end into my nostril and vigorously squeezed the orange bulb attached to it. Surprised at my startled reaction that nearly tipped over the stool, he grasped my arm and squinted at me determinedly behind dirty lenses. But before he advised me he "would be blowing the other nostril," I slipped from the stool and fled down the stairs, leaving Cole to settle the bill.

Another day, the Peace Corps doctor looked up from the slide he had been examining under the microscope and reprimanded Cole and me for eating "contaminated" food at village tea stalls. Jokingly referring to me as a "walking parasite," he opened his medicine cabinet and handed me boxes of tablets for the giardia, amoebas and assorted worms he said I had. That night, Cole and I found a room in a guesthouse in Richmond Town owned and managed by an Italian woman who rented beds for a nominal sum. Nauseous and disoriented an hour or two after swallowing a handful of pills, I tossed my small bag on the faded carpet of the room and started the first of many trips to the squat latrine, where I passed stool after runny stool. Laughing uncontrollably, I vomited thin yellow mucus again and again, until I collapsed weakly on the bed. Alarmed at my repeated dry heaves, Cole called a taxi and rushed me through Bangalore's quiet streets to the Peace Corps doctor's residence.

The recent Harvard medical school graduate took one look at the condition I was in and jabbed a needle into my arm. That night and the next day, colored patterns put me in a dreamy state, until I finally drifted toward the sharper lines of reality to watch Cole's tired eyes smile and an Indian doctor say, "Dr. Smith has been called elsewhere."

Curious about the collection of containers stacked on the bedside table, Dr. Jacobs muttered, "What's all of this?" before he advised me not to take atabrin, chloroquine, diodaquin and "what else do you have here?—teramycin?"—all at once. "We will take care of the parasites in your system one medication at a time," the kindly Indian doctor said, dropping all but one box of the tablets into his bag. Responding weakly that I had only been following the Peace Corps doctor's instructions, I told him I preferred natural remedies; that Mother had roasted pumpkin seeds, then ground and mixed them with brown sugar into a lump the size of a tennis ball. Finding the doctor attentive to what I had to say, I informed him Mother had given the crushed seeds to her patients to be eaten throughout the day with as much water they wanted to rid their systems of round worms.

Knowingly nodding his head when I told him that in the early years she had used neem leaves as a poultice, Dr. Jacobs bid me follow her advice. And, as recorded in *Sepia Prints*, during the Second World War, grandmother dried the bark of the cinchona tree, then ground it into powder to give to those suffering from malaria fever.

At the end of Cole's and my two-year service with the Peace Corps, we were extended for a third productive year in Mandya as volunteer leaders with grants we received from Oxfam, CARE and the Ford Foundation. But before we began our third year, we were entitled to a one-month return to the States, and boarded a train for Madras. At Meenambakkam Airport, the attendant was indifferent to our and others' distress, and repeated what the voice had crackled over the loudspeaker: "I regret to inform you that your flight has been further delayed."

Resuming her duties, the attendant reluctantly dealt with the irritable businessman who shouted at her. Further embarrassing the ticket personnel, the unhappy traveler informed them that the next time he would take first-class rail rather than risk India's domestic airline. Strikes, hijackings and numerous crashes had taken their toll on passengers' confidence, and the airline had earned the title "Flying Coffins."

Motioning for me to follow her behind curtains, the female security guard ran her hands over my body to check for concealed weapons. I loathed flying IA—or any airlines, for that matter, in which passengers met knees to seat backs in discomfort, in uncomfortable contrast to the steam ships on which I had sailed with my family as a child: languorous boat journeys that had taken months to sail past Lebanon and on through the Suez Canal, the Strait of Ormuz, into the Persian Gulf to the Arabian Sea and finally to Bombay.

With hours to wait before the international flight took us far from what we had come to know, we found empty seats near a wilted plant and munched on oily *vadai* Cole purchased at a tea stall. Nearby, an inspector harassed a group of sunburned Italians who had portable radios packed in their luggage. "Batteries?" he queried the tourists, who were eager to comply with his demands but became frustrated when they were sent to yet another counter for further questioning. Sympathetic, I was reminded of what Kipling wrote in the year 1892: "It is not wise for the Christian's health to hustle the Aryan brown," having tolerated much of the same at customs clearance when we tried to convince the authorities we had no intention of selling the radio we carried.

In time, we were told to board the plane, and dozed until our arrival in Aden.

Not far from the airport, soldiers stopped the taxi driver at a checkpoint surrounded by barbed wire. "Surely you're aware that there is an insurgency going on?" the British officer warned, directing his comments toward Cole. But without waiting for an answer, he dismissed us with a wave of his hand towards the Crater's deceptively quiet streets. Checking into a modest hotel, we locked our suitcases in the room we were to occupy and made our way down the stairs to visit the Indian trader we knew lived nearby. Greeting us warmly, Mr. Srinivas was leading the way onto a wide verandah when an explosion ripped away a portion of the hotel's entrance we had only minutes before walked through. Stunned, we turned to our host, who shook his head and said with a heavy accent, "Such occurrences are not uncommon."

Troubled, we left the porch to enter a dark room, where our host shrugged his shoulders and told us of other unsettling incidents. "My wife is not able to join me but has chosen to remain in India," he admitted. Supportive of her choice, I could not have known then that one day we too would live in the midst of combat.

The Lloyd Trestino liner Cole and I boarded was one of the last ships to sail from Aden through the Suez Canal before the Israeli Egyptian war of 1967. Warm days and starry nights later, we docked at Brindisi and boarded a train that took us to the interior of Italy. Thick fog covered the aerodrome in Rome the morning the pilot accelerated our plane for take-off, when, with a terrible lurch, the Boeing 707 collapsed on the runway. Sick with fear, we attended to the steward who had collapsed in the aisle near us. Loosening his clothing, we fanned his face and waited long anxious moments until instructed to leave the aircraft and report to the airport dining room. There the staff offered us wine and pasta before we boarded another Alitalia plane with a great deal of trepidation and an itinerary that included New York, Cincinnati, Lexington, St. Louis, Kansas City, Wichita, Denver, Salt Lake City, Boise, Spokane, San Francisco, Hawaii, Hong Kong, Singapore, Madras . . .

Thousands of miles later, red-eyed, we fastened our seat belts to watch the city of Madras twinkle below on a sultry, hazy night in March of 1967.

"*Taxi sari illah, ayah,*" Cole complained to the driver who parked his rundown vehicle near the airport curb. But it was well past midnight and, beyond the point of caring, we watched the driver adjust the rope he held to keep his door closed before he drove us onto the deserted street. Soon after, the taxi jolted to a halt when the front wheel came off and wobbled down Mount Road. Laughing heartily in spite of the situation, we pulled our luggage from the tired old vehicle to sit on the side of the road until another taxi came along.

On a later January afternoon in 1968, Cole and I formed part of the tourist crowd at the *khedda* round-up that took place sixty miles from Mysore City in the State Forest of Kakanakote. Advised to remain behind heavily roped barriers, we watched men herd dozens of wild elephants down the Kabbani River. Trumpeting in terror, the great anxious beasts crashed through the underbrush ahead of their relentless pursuers, who rode *kumkies*, tame elephants, and brandished sticks. Exhausted, the terrified animals were driven into massive stockades constructed with teak logs surrounded by deep, wide ditches. My heart bled for the elephants that would be subjected to months of confinement before they were trained to work for man.

Such was the fate of one elephant chained to a stake in an enclosure in Bundipur forest, Mysore. Trained to obey the *mahout's* command to kneel, the elephant allowed us to clamber up her leg to sit on the platform strapped to her back. Then, heaving herself into an upright position, she raised her trunk to tear leaves from branches as she lumbered down the forest path. Fond of the elephant he had helped train, the *mahout* informed us she consumed up to 150 kilograms of vegetable matter a day.

"From a habitat that is almost no longer there," he added, which furthered our concern that so much of the world's wildlife, including elephants, continued to compete for the ever-burgeoning human population's demand for agricultural and settlement land. As Dr. C. Santiapillai and Peter Jackson noted in the *Oman Daily* of 13 August 1991, "the Asian elephant is in grave danger. (With) the surviving population of between 30,000 and 55,000 elephants in the wild—only a tenth of that of the African elephant."

The British defeated Tipu Sultan in 1799, but before they did, the "Tiger of Mysore" built a lovely summer palace in Srirangapatnam, where Cole and I had picnics in the beautiful manicured gardens. Farther down the picturesque gravel road, the twisted branches of a banyan tree provided shade near the muddy waters of the Kaveri River that flowed, sometimes swift, sometimes slow, depending on the fall of the monsoon rains. On occasion, we drove to Mysore City, known for its sandalwood products, silk saris and the yearly Dasara festival, to visit our old friend who lived at the edge of the city in a humble home for the aged.

Father Percy's face brightened when we walked into the room. Tearing up at the sight of the eighty-year-old priest who had all his life interpreted believing into caring and now sat quietly near the window listening to birdsong, I gave him a hug and he patted his bed on which we were to sit. Apologizing for his "unfortunate" habit, he opened a small tin box he kept on the table and lit the butt of an old cigar. Holding it between trembling fingers stained yellow from snuff, he puffed slowly and said, "Your coming is worthy of celebration. You know, I so often recall my visit to your house and the wonderful dreams I had sleeping under the mosquito netting you made from your wedding veil," he chuckled. Suddenly despondent, our old friend rubbed the glow from his cigar and stuffed what was left of the stub into his worn box.

"Who do you think is influenced by what I say anymore?" he said with a sigh. "My Indian brothers have long ago taken over what the Lord led me to begin," he reflected, absorbed in his own thoughts. Then he admitted, "You know it troubled me—it still does. That the position of caste interferes with decision-making." In the epilogue of *Sepia Prints*, my brother wrote: "Missionaries came with New Wine, wine they believed could never be fully contained within the caste and other structures of Hindu Civilization."

"Overall," Fr. Percy concluded with a sigh, "we did all agree on the direction of the heart where love transcends."

Often on a late afternoon we walked down the dirt road from our house to visit with friends who lived nearby. One evening, Seshadri Iyer asked us to join him on their verandah and I left the men to join his wife, Padma, in her spotless kitchen. Crouching near her (as I had so often done as a child with my *ayah* in her kitchen) below the polished brass *dekshis* she had lined up on the shelf, I watched her stir foaming milk she mixed with tea leaves in a clay pot. Carefully pouring the steaming liquid into stainless steel mugs, Padma offered the *chai* to the men and me, along with the sweets she had made. Then, leaving us, despite my encouraging her to stay, she took along with her the observations she had so easily expressed to me in her kitchen. Watching her retreat, I reflected on the position society imposed on her and other of my Indian women friends, then turned my attention to Seshadri, who had not faltered in his discourse concerning social and cultural advantages.

"Should we assume the West has the answers—apart from our ancient traditions?" he asked quietly. Gentle with us, for we were much younger than he was, the professor took pleasure in encouraging the exchange of ideas. In response, Cole's expression tightened with sincerity. His words that he was of a different opinion echoed my sentiment and spread a smile over our friend's face. For, although John F. Kennedy had inspired us as volunteers to offer our best, we knew we had a great deal to learn and that the interchange of ideas was of utmost importance.

"I suppose much of my motivation for joining the Peace Corps was really from a desire to return to India," I conjectured, and went on to surmise that I thought Cole's was partly to leave roots he was certain did not provide the answers; for he had come from a community where the kids were channeled into the trades, as academia was thought to have few job prospects.

That evening, as the last rays of the sinking sun silhouetted the distant clouds, Cole and I sat on the banks of the irrigation canal and mulled over what we had discussed with our friend.

"My motives are not really so altruistic as they are just to make it," Cole commented, throwing a stick into the swiftly moving water. It tipped crazily and bobbed this way and that before becoming part of the current . . . not unlike the course our lives would take.

Our interaction with those living at the training center and in the surrounding villages taught us fundamental lessons concerning development with a difference that served us well over the years. Village life broadened our ethnocentric perspectives, equipping us with added cross-cultural communication skills that helped us appreciate other ways of being in the world. Working with supportive colleagues, Cole and I formed a strong bond of trust that year. Despite the fact that our experiences were of far greater value to us than the transfer of our skills to the people with whom we worked, our involved interaction helped promote what we aspired to achieve—the art of peace.

On completion of our three years with the Peace Corps, Cole was keen to further his studies and took up our friend Ed Harper's (who had been conducting research in a village nearby) suggestion to study anthropology with him at the University of Washington. We arrived in Seattle in 1968, at a time when students were venting their anger against the Vietnam War. We too were against the war, and after I finished my work at the University Hospital, and while Cole attended classes, I joined the demonstrators covering six lanes of traffic to sing: "All we are saying is give peace a chance." For, as the poet Wilfred Owen said, "If we saw the pain and suffering of war, we could not tell the old lie that it is glorious to die fighting . . . the face of the enemy is not very different from yours."

Cole then applied as a conscientious objector in opposition to the unpopular war, and was ranked number 120 in the military draft lottery system. Encouraged with the support he received from the Quakers in Seattle and the lawyers of the American Civil Liberty Union, he was nevertheless turned down by the Spokane draft board. But after appealing again, he was granted status as a CO and we began his alternative service with Quaker Peace and Service in Nigeria, a country that had just experienced the Biafran Civil War.

Chapter III
Post-war

FOR THIRTY MONTHS, the Ibos were packed into forests and swamps that could not support them," David Lamb writes in *The Africans*. They "began starving to death at the rate of a thousand a day," until January 1970, when the Ibos surrendered and Ojukwa went into exile. Many of those who came to help those affected predicted widespread reprisals against the Ibo, both by the federal troops and other minorities in southeastern Nigeria. The fundraising and media statements that reflected support for Biafra blocked benefactors' chances to work in the area once the war was over. However, the Quakers, who had condemned both sides of the conflict with pacifist passion, were able to merge their personnel from the federal and the Biafran sides of the war and transition from relief to rehabilitation.

Believing reprisals could quite possibly accompany peace, the Quakers believed their presence would help lessen the revenge motive. They also possessed a small pot of money left over from the emergency, as well as committed supporters, doctors and nurses who had served in the relief operation. They were spurred to commitment by the daily horror they witnessed and continued to have a personal interest in the aftermath of the conflict. Finally, the Americans, through the Agency for International Development (AID), and British governments offered to allocate funds to Quaker Peace and Service in order to keep up the international pressure amid tribal rivalry. The two million dollar, two-year Quaker program we worked with included the rehabilitation of 210 villages with community centers such as schools, dispensaries and churches; the mission hospital in Abiriba; and the rebuilding of Maternal Child Health facilities in and around Uyo, Ikot Ekpene and Itu; as well as the renovation of the Ibusa and Asaba town water supplies. Finally, sixteen thousand children housed in thirty-five orphanages were reunited with their families through the dedicated persistence of social worker Rhoda Veni, who photographed every child and gathered as many case histories as the caretakers and children could provide. Her admirable pursuit resulted in only one of the orphanages being left open to cater to the handful of children left without families.

Most mornings, over bread, fruit and tea, the Quaker team, of which we were now a part, divided up duties and deliveries. Decisions were reached only after the most doubtful participant stated his or her undisputed position. This left the rest of us modifying our opinions. This process was sound in theoretical terms, but consensus decision-making and course corrections left the program circling. As the logistics officer, Cole raised the issue of transportation, with the intention of sorting out who went where, when and with what Land Rover, Bedford truck or Peugeot 404, only one or two of which were capable of long-haul trips to Aba, Port Harcourt or Lagos. Loyal drivers continued to sort out most arguments, after which we often met up with each other in remote villages or passed one another on tracks that led through thick forests where vines shrouded rusted vehicles left from the recent civil war. The African staff fit right into the melee of things, with the exception that they could return to their families at the end of the day. They did not live communally as we did.

Lantern light encircled those of us around the table one night as we shared a calabash of palm wine. "We have found a room to rent just a ten-minute walk away," Cole told our Quaker colleagues. The next day, pygmy goats scampered ahead of us as we carried our belongings to the home of our Ikot Usen friends, who welcomed us to the small room connected to the side of their mud house. Alert to our needs, our host dug a pit latrine for our use and skillfully thatched an area where we could wash ourselves. Most evenings, I prepared a rice dish on the kerosene burner, but we welcomed invitations to share *garri*, cassava porridge, and rich fish or *igussi*, gourd-seed soup, with the family, who became our friends. Those evenings, we sat on the sun-baked earth and chewed bitter kola nut until moonlight spilled on the tranquil scene and time flowed easily into night.

One muggy morning, haze covered the little village of Itu. At the bottom of the hill, the Cross River flowed with brown silt through the valley. Although it was still early, men shiny with sweat bailed water from the poorly patched, bullet-perforated boat with a rusty can. Swallowing my anxiety, I watched the boatman load the last of several jerry cans filled with petrol for the twelve-horse-power engine before he pushed our overloaded, leaky boat away from the bank with a heavy bamboo pole.

The Cross River swirled—a mile or more across in places. Undergrowth flanked the river alongside banana trees and ferns that stretched their fronds toward the treacherous current. Post-flood debris swept downstream and our companions related how they had seen bloated corpses carried down the river during the war. Following an hour or two of blistering sun reflected from the water, the boatman maneuvered the boat to the bank, where we pulled green oranges from low branches. The tangy fruit refreshed us until the drone of the engine was interrupted by Afiong's shouts. Full of life, her hair braided into spiky antennas, she caught the rope the boat driver threw her and pulled us ashore to Uman Island. Offering us steaming cups of milk, for there was no tea or coffee in the market, she led the way to where villagers who had lost their homes to floodwaters had erected temporary huts. Later, while rain fell heavily on the tin roof, villagers' voices flared into tempers with those who expressed competing demands for assistance.

Afiong was the first to jump from the boat at Ikot Okpora. Sylvester and his wife were generous with their hospitality and served us peppery hot catfish cooked in palm oil. Mosquitoes swarmed that night and before the first cock crowed, I made my way to the cooking fire where, oblivious to the smoke, I crouched near the flames to avoid the sand flies that left angry welts on my arms. Breakfast of *garri* and leftover fish soup followed, after which our hosts gave us enough bread and bananas to last the five-and-a-half-hour boat trip.

An expectant line of villagers waved their arms to welcome us on the shores of Upper Biase near Okurike. Leading us up the muddy incline to the chief's hundred-year-old house, the occupants insisted we have something to eat before we discussed relief assistance. Later, in the close confines of the village church, restless children fidgeted at my feet and I struggled to keep attentive during their lengthy speeches. Suddenly, wind slammed the shutters of the window and an approaching storm darkened the room. Summoning her little ones, a young mother left the gathering as thunder clapped and lightning split the sky.

Despite heavy rain, we slipped down the embankment to the boat to find our fellow passengers already grouped under umbrellas, plastic sheeting and palm fronds. Then, as the sun slipped into the river, our friends' faces were silhouetted against the deep crimson sky as they harmonized their voices in song for the hour or two it took to reach Ikot Okpora. Our hostess was pleased with the wild boar we purchased for fifteen pounds. She simmered the meat with palm oil and hot peppers and served it with yams.

The following morning, the sun shone white-hot as the current took us swiftly downstream along banks where heavy shelling had resulted in the loss of many lives. Near Itu, small boys jumped into the river and pulled us ashore to mark the end of our voyage on the Cross River.

"It's always a bit dangerous downriver and thereabouts," we were told on our return. Those who advised us traveled with armed escorts the federal government offered the Quaker staff. Concerned that her husband had not returned from Itu, Effiong Etuk's wife, Stella, watched the clock tick past midnight before she convinced Cole to drive her to check on his whereabouts. Unable to find the men (who later did return), Cole had started the drive back to Ikot Usen when he noticed villagers blocking the road ahead. Swarms of insects covered the headlights when he slowed down to find men holding their comrade, who had been shot at close range. The wound was severe. Wiping away as much dirt as he could, Cole loaded the injured man and his friend into the car. He then coaxed the Land Rover up a steep, muddy hill to where he knew the German Emergency Doctors were located.

"Better take him back to the village to die," the doctors advised, after examining the injury. But Cole insisted they clean and stitch the wound, and was later pleased to hear the man had fully recovered.

Our experiences in the rural setting came to a close when Cole, at the age of twenty-four, was designated the Director for Quaker Peace and Service for Nigeria. Boarding a plane for Lagos, we unpacked our suitcases in a dreary two-story concrete building situated above the office at 16 Sumbo Jibowu. When the cook we inherited from our predecessors admitted he would have quite happily thrown the wife of his previous employer over the railing, Akpan Udon, who cleaned the premises, turned his face to smile.

Soon after our arrival, one of the 165 nationals Cole administered drove us through the back streets of Lagos to watch Fela Ranson Kuti, "007," perform. Warmed by beer and dressed in colorful attire, the appreciative audience moved their bodies and tapped their feet to the captivating beat. Overcome by the congestion and stifling heat, Cole and I left the crowded room and found a place to sit under African skies listening to the throb of drums. "It's going to be a real challenge to deal with personnel," he sighed. And indeed, one of them later suffered a nervous breakdown and had to be evacuated.

The next afternoon, we took a walk past open drains to the city market. Absorbed in selecting a piece of bright tie-dye material from the stack of fabric for sale at a small stall, it was not until I heard loud laughter that I noticed Cole surrounded by a group of spirited women. And as I watched, a Yoruba woman clad in traditional cloth swayed her generous hips toward the animated group and snatched the pens and notepaper from Cole's shirt pocket. Tucking the loot into the skimpy brassiere covering her ample bosom, she enjoyed his confusion and chuckled seductively, saying, "Come and get-o," before, arms akimbo, she retraced her steps and handed him his belongings.

"What a contrast Yoruba matriarchs are to the women we knew in Asia," Cole remarked as we continued walking through the market past piles of dried fish—the popular ingredient Nigerians use in soups, which are eaten with yams or cassava. Ahead on the path, feral cats spat and fought over fish bones. Clapping them aside, a man dressed in blue embroidered robes beckoned to us from behind a table stacked with withered dried bark, various herbs and skulls of monkeys: items he said were "good for manhood." Smiling, we walked on by.

Another day, we closed the doors of the Quaker office early in order to take my brother and family to Bar Beach and the pleasant respite of salt air. John and his wife had left Kodaikanal where they taught, to sail across the Indian Ocean and pay us a visit. Suddenly, with a shriek, a man with braided hair and painted eyes leapt onto the bonnet of the Peugeot Cole was driving. Drumming frenziedly, he gyrated avidly, as did men on other cars. Confusion reigned, and agitated drivers honked and shouted at the ecstatic dancers. Just as suddenly as they had appeared, the wild apparitions bounded away and the spell that had disturbed our senses was broken—allowing us to drive on toward the crashing surf, where Cole parked the car near a small wooden stand.

"Me, I am a celebrity now!" our Yoruba friend exclaimed as she walked toward us. Helping her ease the basket of fruit from her head to the warm sand, I watched her slice a succulent pineapple before she wiped her hands on the cloth wound about her waist and showed me a glossy picture of herself on the cover of a recent *Span* magazine. Justifiably proud, she told me she had sold enough produce to afford to pay for her son's medical studies in the U.K. "He would like me to visit him, but my place is here," she told me we sat together watching huge waves surge Cole high into the air. Later, I joined him in the foaming waters of the Atlantic Ocean that tossed us about like corks.

But Bar Beach had its dark, sordid side. Blood spilled on the white sand when armed robbers were publicly executed. Offered free transport to witness the gruesome events, Lagos residents became victims themselves when men picked their pockets while they watched the brutal proceedings.

One later evening, in the courtyard of the American school where I taught music, the little son of a Nigerian politician solemnly sang a song he had helped compose. He was standing near his father, who had taken part in authorizing the recent executions on Bar Beach.

Lighting the last candle, held by the youngest in line, I walked with the children to where their parents were seated. The children sang in sweet unison songs of their own creation about the benefits peace could bring. And as the politician proudly watched his son, I studied his shadowed face and wondered whether he chose not to recognize what Henry David Thoreau suggested was every man's task "to make his life, even in its details, worthy of the contemplation of his most elevated and critical hour."

Toward the end of our second year in Nigeria, Derek Robinson, whose job with Oxfam in Togo was to assess and recommend projects in West Africa, invited us to drive across the border to discuss Cole's request for a substantial grant to pay salaries, rent and other recurrent costs for the project he had been working with. Subsequent discussions with Derek regarding long-term commitments to rehabilitation and development set in motion our own future employment with Oxfam's well-developed involvement.

Less than two years after we had arrived in Nigeria, Cole and I took down the Michelin map of Africa that covered the wall of our front room and left behind three Quaker Service doctors employed by the Ministry of Health to complete the work we had begun. Certainly the job in Lagos had not been easy, but administering the staff at the Quaker office had given Cole personnel experience that would be useful to him in other positions, although the number of workers he managed was far below those he would supervise in the countries we were yet to live in. We were soon to embark on the much more challenging, stimulating and interesting job of parenting. For, within a space of less than a year, we would have three children under the age of three.

Chapter IV
And Then We Were Four

Jessica

JUST DAYS AFTER Cole had completed his masters in anthropology from the University of California, Berkeley, where we had been living for nine months, we hugged Jessica, our little two-month-old Kiowa Indian–French-Canadian daughter close, and celebrated the completion of the Oakland, California adoption formalities at our favorite Japanese restaurant. Devoted and anxious as parents, we rocked Jessica Shanti Ko-sahn in the wooden cradle Cole made for her, and when she cried, we tucked her under the blankets near us. When time permitted, we traveled the coastal roads that wound near spectacular views and walked for miles in the bracing salt air with Jessica wrapped tightly on Cole's back.

Then, when Cole was offered a position with Oxfam America in 1973, we sold our cottage nestled in the red woods for what we had paid for it. Packing our possessions into a U-Haul truck, we drove across the country to Boston. Throughout that year and the next, Cole generated income for the bankrupt American arm of one of the world's best charities through direct-mail fundraising. He frequently appeared on television talk shows and talked to students, businessmen and churchwomen's guilds in order to earn enough money to enable the organization to move from a damp, crowded church basement to a building in south Boston. In time, we wearied of hectic schedules, and since Cole's position had been offered to the brother of a board member, we requested a location overseas. Cole was offered the position of Oxfam Field Director in Bangladesh.

It was a relief to step on firm ground after the turbulent flight to

Dhaka, Bangladesh, in August 1974. Sticky heat filled every pore as I struggled across the tarmac with the heavy hand luggage, while keeping an eye on our two-year-old daughter, who lagged behind to splash through dirty puddles.

It had been a month since we had seen Cole, and when I caught sight of him at the airport, I dropped the luggage and waved both arms. Jessica ran to her father, who threw her into the air before he hugged me close and led us toward the immigration desk. Outside the airport, women waited expectantly with their wasted children under a neem tree near the Land Rover. Wretchedly poor, in the most densely populated country in the world, the women's pathetic appearance affected Jessica, who handed the biscuit she had been eating to a child. Cole pulled *taka* from his pocket and handed the money to the women, who thanked us and extended their hands for more. "That will last them for half a day," I said unhappily, as I watched them walk off counting what they had received.

"Difficult times for Bangladesh," Cole said as he started up the Land Rover. Certainly, what I had already observed was not what I had known in India. Yet "Bangladesh is uniquely successful," as Clarence Maloney writes in *Behavior and Poverty in Bangladesh*, "with the intensity of human interaction, the expressive culture . . . remarkable." Maloney goes on to say that there is probably no other society in the world in which such a heavy population could subsist without destroying their deltaic land—unlike industrialized societies that, with far less population density, use up their resources within a few decades.

Surely not, I hoped, when Cole turned off Mirpur Road at a major junction and parked near the single-story cement structure that would be our home. But home it was, and that night and during the following weeks, the aggravation of traffic sounds, shrill air horns and conductors raising their voices to persuade passengers to board buses bound for Savar and Tangail further encouraged us to move our office-cum-residence to a quieter side of town. Just before we were scheduled to shift to Road 21, in Dhanmondi, Cole traveled to Comilla to speak to a group of NGO leaders. Discovering that his return flight was cancelled due to a disruptive *hartal* strike, he chartered a three-wheeled auto-rickshaw to take him to the first of five ferry *ghat*s he crossed between Comilla and Dhaka before he took whatever transport he could find to arrive home. Only to announce the following day he would be traveling to the extreme northeast of the country to an area sandwiched between the mighty Brahmaputra and the Indian border, with no road access at all.

Assuming train travel in Bangladesh to be similar to what we had experienced in India, Cole purchased a first-class ticket. Arriving at the station, he found the four-berth compartment already occupied by five other men, all of whom went to the conductor to pose their complaint.

"If you are not liking the arrangement, sir, simply give me the tickets and I will be selling them to others who wish to travel," the man said, shrugging his shoulders. This was something of a falsehood, as many who traveled simply did not purchase tickets.

Taking no chances, as the rail line to Mymensing was notorious for *dacoitry*, the men stacked their luggage against the door that had no lock and agreed to take turns to keep an eye out for intruders. Cole crawled up on an upper berth and quickly fell asleep with the motion of the train, only to be awakened in the dead of night by the weak light of a torch that revealed a knife held to his throat.

"Money," the *dacoit* hissed.

Struggling to a sitting position, Cole pulled out what he had from his pocket, having hidden the rest of his cash under the seat. His comrades were not so lucky. Skirmishes followed. One of the men was stabbed in the stomach and all of them were stripped of their watches and clothing. The *dacoits* then ordered the men to stay in the latrine, while they lit a small fire in the compartment (the light bulbs had long ago been pillaged) to see what valuables they had accumulated. Wrapping their plunder in a shirt, they jumped from the slow-moving train and Cole reached for the brake chain, which eventually stopped the train. The conductor pushed through the crowd of excited commuters who surrounded the compartment. Nodding his head sympathetically, he said, "Theft is a common occurrence. May Allah be praised that you have not lost your lives."

Early the next morning, hundreds of passengers scrambled off the roof of the train when it halted at Bahadarabhat *ghat*. Having had his shirt and pants stolen by the *dacoits*, Cole purchased a *lungi* wrap and *kurta*, shirt, before he boarded a small country boat towed upstream with long ropes by two men walking along the banks. Well into the day, the boat man blew on the embers of a small fire and prepared *dhal bhat* for their late-afternoon meal. Intermittently, showers washed over the little boat and left the travelers damp as they sailed the thirty-six hours down the river to Rowmari. The river collected melted snow from the Himalayas before the water flowed through the tributary systems that during good years spread rich silt on the *char* land where crops were planted.

But other years, floodwaters covered the Bangladeshi lowlands, accompanied by cyclonic tidal waves that caused terrible suffering, famine and loss of life. Ten of the century's deadliest storms spent their destruction at the northern end of the Bay of Bengal. The worst of these disasters was on 13 November 1970, when Bangladesh was still known as East Pakistan. Thousands of survivors blamed the government in West Pakistan for not warning them in time or providing relief after the devastation. Their anger helped fuel a civil war, and Bangladesh declared its independence from Pakistan thirteen months later.

The floods of 1974 caused a major famine in Rowmari, forcing destitute rural families from across the country to abandon their rural homes for towns, especially Dhaka, where they joined the already desperately poor "*bustee* camp" dwellers on whatever open land they could find.

Distressed by the appalling situation in Rowmari, Cole abandoned the idea of returning to Dhaka on a towed country boat and persuaded a driver with an engine-powered boat to speed him downriver to catch up with the passenger train bound for Dhaka so he could alert his Oxfam UK colleagues about the crisis. The next morning, he arrived home weary, bedraggled and clad only in a dirty *lungi*. Smiling at my obvious pleasure when I opened the front door, he gave me a hug and leaned down to pick up Jessica, who had hidden behind my legs until she recognized her Papa. Thereafter, she refused to let him out of her sight and accompanied him to his meeting, at which he consulted with his colleagues before he telexed headquarters in Oxford, who confused the quantity of forty-four thousand pounds he had requested in seed weight with pounds sterling. This was a large amount of money, considering Cole had inherited an annual program budget of $100,000, with hardly enough to respond to the emergency needs everywhere apparent.

Smallpox had spread through the *bustee* camps, raising fears that the global eradication campaign would fail. Cole made emergency grants in support of the smallpox eradication program led by the World Health Organization (WHO), and funded F. H. Abed's new nongovernmental organization (NGO), Bangladesh Rural Advancement Committee (BRAC), to undertake a massive famine-relief program in the north, while simultaneously purchasing winter wheat seeds from India for experimental planting under the Mennonite Central Committee (MCC) program.

This was a busy and meaningful time for Cole, who transitioned the Oxfam program to a large international NGO with CARE and Catholic Relief Services to new dynamic directions in support of BRAC and Gono Shastra Kendra, founded by the charismatic Dr. Zaffrullah Chowdhury; Gono Unnayan Prochestra, the old Quaker Peace and Service project in Rajor, Faridpur and Kumudini; and other projects. These provided him with excellent contacts when he later served as the UNICEF Representative for Bangladesh from 1989 to 1992. Those years he "decorated many chairs," and UNICEF spent $30,000,000 annually for sponsored projects that produced tremendous achievements in child immunization and sanitation—schemes that were encouraged by an enthusiastic government.

Worth a story in and of itself, "Samantha," the Land Rover Cole drove to Rowmari, had originally been owned by the Liverpool Fire Department before it was air-freighted to Iran, then Peru, for earthquake relief. It was thereafter used in another eleven countries around the world before Operation Omega volunteers drove Samantha from London to Dhaka in 1970.

Cole drove Samantha straight north to an area that had not been driven to since partition, when Rowmari had been accessible from Assam. Waving at the nervous troops he confronted at the Indian outpost, who hit their bunkers and pointed their guns, Cole turned south on a bullock track that cut across dry rice paddies until he eventually came to a branch of the Brahmaputra (known as Jamuna in Bangladesh) River, where he waded across to test the knee-deep water. Halfway across the river, the current washed sand from under Samantha's wheels and the engine stopped. Hurriedly twisting the tire iron around the fan belt until it broke, Cole sloshed back to jump behind the wheel. Fortunately, the engine started and didn't stop until he reached the far bank, where he put on a new fan belt. Thereafter, he used this technique when crossing the river on his frequent trips.

Months later, Cole, accompanied by his photographer companion, Nick Fogden, drove Samantha onto a ferry that ground to a halt on a sandbar. Not one to be easily deterred, Cole backed the Land Rover off the barge and drove up the embankment onto the rail-track bridge that spanned the river. Suddenly, he saw a train loom ahead. Frantically flashing his lights, he breathed a sigh of relief when the engine screeched to a halt, and he continued to bump the vehicle over the ties to the end of the bridge. Coaxing the Land Rover down a steep slope, he parked near a pier and found a boat driver willing to speed them down the river toward Rowmari. Still with fifteen kilometers to walk, Cole and Nick took time to photograph and speak to those administering the relief operation, who were feeding fifteen thousand malnourished children. By the time they arrived back at the boat, the sun had set.

"Too late, we are leaving too late," the boat driver complained, before starting up the engine. Soon after, the boat skimmed onto a shallow mud bank and the propeller blades whipped black masks on his passenger's faces. Laughing heartily, knee-deep in mud, the men pushed the boat back into the water, only to lose their way repeatedly in and out of false inlets. Relying on the moon full with silver light, the boat driver raced through a channel of water beside a twelve-foot embankment, too late to avoid the raft made from the stalks of banana trees. "Allllllllllaaaaaaaaaaaaaaaah," the boat driver breathed as he and his passengers looked back to see the survivor of the banana boat bobbing in the water waving his fists.

"He is all right, thanks be to God, but *khub khub* angry" the boatman reported, shaking his head as they proceeded down the river, slower now. The men were an interesting sight when they reached home the next morning.

Sanjeeb

It was in Bangladesh that we found our son.

"I've just completed the paperwork for a two-and-a-half-year-old little boy God must have intended for you," Sister Priscilla told Cole. He had told her of our adopted daughter and presented her with a check for $5000 a friend in Boston had given us to donate to Mother Teresa's Sisters of Charity home in Dhaka.

The next day, sweat coursed down Cole's handsome face as he drove Jessica and me to Mother Teresa's home in the old city. During Moghul times, the narrow lane we were driving on had been a significant trading center, with streets named after crafts or artisans. Among these were Joginagar, where fine muslin had been woven, and Shakhari Bazaar, where, to this day, *shankha*s, conch shell bracelets, are made. Throngs of pedestrians walked alongside gaunt-faced men who rang their bells incessantly as they wheeled bicycle rickshaws packed with passengers down the crowded Islampur Road. Overhead, crows flapped their wings and swooped near merchants who sold brass and pink pearls in small shops. Beggars and homeless people sat alongside the road. One soul with matted hair stood alone, contemplating yet another day and where his next meal would come from, reminding me of M.R. Rousseau's article "Yearning for Nightmares," in which he told of a rich woman who, having had a nightmare about her child, presented a destitute man with a live chicken. His response was to say he would pray for the woman's child, with hope the woman would have many more nightmares!

The *chowkidar* who guarded the entrance of the Shishu Bhavan compound shouted at the barefoot children gathered around the man near the gate, who pushed a cart piled with puffed rice mixed with green chilies and mustard oil. Purchasing paper cones of rice for the children (of course it was not enough), we were bothered by those who jumped on the fenders of the Land Rover asking for more. Irritated, the watchman scolded the children and shooed them away before he opened the gate to the Missionaries of Charity.

Inside the compound, a sudden hush prevailed, as though we had entered another world. High, thick walls closed out the din from the street. Even Jessica was quiet as a sister led us to a dark room where Sister Priscilla was seated at a desk under a whirring fan. Rising to greet us, she recognized Cole and suggested we follow her up the stairs. She led us to a large room, with row upon row of cribs of every size holding babies. Some of them slept, while others lay quietly with wide, staring eyes.

Near the window, an *ayah* held a little boy clad only in diapers. Having been told the reason for our visit, she pointed to me and I held out my arms for the little one.

Pleased with the child's response to me, Sister Priscilla, whose face showed weariness, raised her arms in prayerful thanks toward the flaking paint and brown stains of the cracked ceiling.

"The structure we are standing in is very old," she said with a sigh. "Every year, heavy monsoons threaten Dhaka's old buildings. Not long ago, in pouring rain, the *ayah*s had just finished washing the bowls from the evening's meal of rice porridge in the room located below the dispensary, when the room attached to the one we are in—look, just through that door,"—she pointed to the door leading into the next room, a few feet from where we stood—"collapsed behind them in a terrible crumbling heap."

Staring at her in disbelief, I said, "Just there?" and added anxiously, "Did anyone die?"

Nodding soberly, Sister Priscilla told me two children had perished under the debris of churned plaster, dust and metal beams. "But then," she added, hugging the child next to her, "with praise, we thanked God for protecting the rest of us."

"What about this building?" I inquired, trying to visualize the tragic incident.

"Is it safe?"

But before she could answer, a baby cried, and I went to comfort the little girl, who flailed her thin arms and met my tear-filled eyes with her dark ones. Unclasping her tiny fingers from mine, I wished those caring for the children well and followed Cole, who carried Sanjeeb Paul down the worn stairs to sign the required adoption papers. Thanking Priscilla, we left with our son, past the guard at the front entrance, who waved both his arms and shouted for the rickshaw *wallahs* to make way before he closed the gate behind us. Sanjeeb, meanwhile, sat quietly on my lap and watched everything intently, for he had never been exposed to life outside the Shishu Bhavan.

"The rickshaws have more people in them than we have in the Land Rover." I said to Cole, who was inching the vehicle through the narrow, teeming street. Certain the rickshaw pullers scarcely earned enough to provide food for their families, after paying a rental fee to their *mahajans*, I went on to remark that at least rickshaws were environmentally friendly and carried people down Dhaka's back streets, which were often far too crowded for other transport. Although there was much talk of ridding the city of these "archaic conveyances," rickshaws provided employment for thousands of men as pullers and for those who painted bright folk art, assembled the bamboo frames and repaired the bicycle parts.

Branching onto the main road, we were warned by rickshaw cyclists who twisted on their seats to shout warnings about angry protestors who had abandoned their strike and were now rioting on the streets. Proceeding cautiously, we drove by shop owners bolting their metal doors as frenzied men smashed windows and overturned vehicles, which they then lit on fire. Avoiding the burning tires that blocked the road ahead, Cole found an alternative route to drive us back to the quiet streets of Dhanmondi.

But just before we reached our home on Road 21, two frail and desperate women stumbled behind a cart filled with bags of wheat. Stretching their tattered saris under a hole in the jute bag, they collected the grain that seeped out until the stronger of the two pushed the other woman aside.

Exhausted from the emotion of the day, I led the children into the bathroom and filled a bucket with warm, sudsy water. Sanjeeb was small enough to climb into the bucket and splashed near the bathtub, where Jessica soaked with sodium bi-carb to relieve her of the heat rash that constantly plagued her. Thoughtfully possessive of her new friend and brother, Jessica watched as I applied gentian violet to help rid our little son of scabies. Our boils, however, persisted long after Sanjeeb Paul's had disappeared.

The children became a common sight dipping into the duck pool and running barefoot through the grass. Mornings, they raced each other on their three-wheeled rickshaws and darted in and out of the garage that served as the office for Oxfam's program. Seated around the table with the rest of us, the children ate their noon meal of *dhal bhat* followed by *mesti dhoi*, sweet curd that we purchased from the corner vendor in small clay pots, accompanied by our company's conversation regarding the effectiveness of anaerobic communal latrines.

Jim Howard, a tall and robust Quaker sanitary engineer, and his two young microbiologist colleagues had been our guests for six months. They were involved with a study at the Cholera Research Hospital and were utilizing two large fuel bag tanks that had formerly been used for jet aviation fuel in Vietnam in which discharged feces from patients at the hospital were stored.

"After ten days, the liquid will become strong ammonia that can then be released into the fields," Jim explained. "Because, during the absence of air in the bags, the pathogens are broken down so that they become inactive, and . . ."

Finding it difficult to swallow a mouthful of yellow lentils after listening to what Jim was describing, I wearied of the daily routine, which included our three guests departing for sanitation research after breakfast before they returned for lunch and dinner to talk pathogens, shigella and E. coli, as well as concerns relating to the thousands of tons of fecal matter released by Bangladesh's population and other (sometimes revolting) topics with whoever joined us for a meal. But many of our guests became very good friends. One afternoon, over biscuits and tea with lemon, as milk was scarce, D. A. Henderson, the head of the worldwide smallpox eradication campaign, warned us that there was still smallpox in the area. For it was not until October of 1975 that the last case of the dreaded disease was discovered in Bangladesh.

Many who came to help accomplished what was beneficial for the destitute, but others did not. One day, I set out to see for myself whether the rumors we had heard were true.

"How long has this baby been here?" I asked the timid girl left to care for the babies at the small orphanage Cole had asked me to investigate before he donated Oxfam funds.

"Five months," she reported.

How could that be? I wondered, concerned because the babies looked lethargic and wasted. Showing me around the premises, the girl-*ayah* proudly explained that the small patch of soya beans they had planted provided soya milk, which they fed the babies. Then the expatriate woman in charge arrived and inquired about my purpose.

"The babies are not well!" I blurted out.

Unperturbed, the young woman told me she was feeding them a nutritive substitute for milk that could be locally grown.

"Just think of the practicality, the money saved," she told me.

Alarmed, I left, but not long after, I returned to find the same woman wrapping a baby in cloth.

And before she could tell me what I already knew, I told the rickshaw driver to take me home, where I threw my arms around our two healthy children and wept . . . not to be consoled until Cole and I reported the woman to an American embassy officer, who deported her and found the babies an alternative home with proper comfort and care.

Chapter V
The Back Streets of Dhanmondi

"By trying, we can easily learn to endure adversity, another man's I mean."
—Mark Twain

IT WAS A stifling afternoon. The phrases I had been memorizing blurred on the page in front of me as I repeated "*Khotai jaben? Eta bas khotai jabe?*" to my Bangla instructor.

Hearing the screen door close, I held out my arms for Jessica and Sanjeeb, who, drowsy from their naps, toddled irritably into the room but brightened at my suggestion of a rickshaw ride. Closing the metal gate behind us, we walked down streets shaded by mango trees to the corner where men raced toward us pushing rickshaws. The puller who arrived first wiped the sweat from his face on the sleeve of his torn shirt before he helped the children onto the rickshaw. Heaving myself onto the seat between them, I was unhappy that a man who was half my size pulled the three of us, and said as much.

"But how will I manage tomorrow if I have no work?" the man said, throwing the words over his shoulder as he strained his thin limbs to pedal us down the street. He and his family had recently lost all their scanty possessions when, during his last months in power, Sheikh Mujib had the *bustee* dwellers forcibly evicted to an area some distance from Dhaka. Referred to as "marginal statistics" in the local press, thousands of slum dwellers, whose existence was already subhuman, were disposed of on resettlement sites. The unfortunate people had no access to water, sanitary facilities, electricity and transport—if they had means—to commute to their daily work in the city. With what little they were able to scavenge from the bulldozed and burned debris, the marginalized people again constructed waist-high hovels in which to ponder the next determining cause of their fate.

In contrast, the week before, we had attended a Bangladeshi wedding where the wealthy hosts had fed a thousand guests dressed in gold and finery a sumptuous feast in their elegant home. At the approximate cost of 100 taka per person, the total came to 100,000 taka, an amount that could have helped thousands of hungry, destitute people.

One morning at the breakfast table, Graham Clugston, an Australian relief worker who lived with us along with his Nepalese wife, reported emphatically, "Over fifty percent of the children are severely malnourished." The results of the nutritional survey he had just completed in the *bustee* camps sent shockwaves through the donor community, and later caused Cole trouble with Bangladesh's head of state. Doubting the accuracy of the nutritional report, Trevor Page, the Director of the World Food Programme, who in the early 1970s made headlines for his management of the relief operation in the Sahel, paid Cole a visit with his own objection.

"Bangladeshis are naturally small," he reminded Cole, and added, "Clugston's report was probably not accurate."

This time, accompanied by representatives from other NGOs, Graham repeated the survey and found the same alarming count. Tragically, the most severely malnourished children were not brought to the free feeding program, but left to die in the dark recesses of their tiny hovels.

Now talk around our dinner table concerned the unhappy consequences of malnutrition and inter-agency rivalry. I said quietly that what was needed was to share with the human family. For, as Mahatma Gandhi reminds us, "The earth provides enough to satisfy every man's need, but not everyone's greed."

Bangladesh Zindabad 1975

Two months after we took Sanjeeb home, on the morning of 15 August 1975, the sound of gunfire woke us, with volley after volley of shots coming from the president's residence near Dhanmondi Lake. Mistakenly concluding that Indians were celebrating their independence day, as it was the 15th of August, we heard the door open.

"There are too many soldiers on the street," our *ayah* complained, wiping the sweat from her face with the end of her sari. Hurriedly switching on the radio, Cole turned up the volume, and we were stunned to hear an army mutineer announce Sheikh Mujibur Rahman had been killed and that martial law had been imposed on the country that hereafter would be known as the Islamic Republic of Bangladesh.

"*Bangabandhu* no more," Veronica whispered softly, as she rocked her body to and fro. "No more Mujib," she said, referring to the Father of the Nation, who had fought for an independent Bangladesh to be the world's 139th country.

Throughout that morning, bursts of machine-gun fire continued to interrupt our peace of mind, until curiosity got the better of Cole and he ventured out onto the deserted streets. Not long after, the scream of MiGs flying low over the city and a truckload of rowdy youth sent him hurrying back to our residence. Later that afternoon, while Sheikh Mujib and many of his family members lay brutally murdered at 32 Dhanmandi, the new President, Khandaker Moshtaque, described the killing of the Sheikh as a "historical necessity," and ended his first speech with, "Bangladesh *Zindabad*!"

"The sycophants needed no encouragement to switch loyalties," Anthony Mascarenhas says in his book *Bangladesh: A Legacy of Blood*. They went to the new president's house "to fawn on Moshtaque and any of the majors they could find." At the High Commission in London, young Bangladeshis stormed the building "and tore down Mujib's photographs and assaulted his personal intelligence officer." Instead of having the boys arrested for trespassing and vandalism, Mascarenhas reports, "the High Commissioner, Syed Abdus Sultan, who had always shown himself greatly devoted to Mujib and was one of his most trust appointees, entertained them to tea in his office after removing the many photographs of Mujib which normally adorned the room."

Years later, in 1989, when Cole was the Representative for UNICEF in Dhaka, the highly volatile Bangladeshis reacted strongly to Ayatollah Khomeini's fatwa against Salman Rushdie. Offering a reward of one million dollars to the writer's executioner, Khomeini promised a place in paradise to anyone killed in the attempt. Rushdie's book, *The Satanic Verses*, was said to be a "blasphemous novel," and, according to the *Bangladesh Observer*, the book was banned in a dozen countries (including Bangladesh) and restricted in seven others. California bookstore owners lamented that because they had displayed the book on their shelves, an atmosphere of "implicit threat" prevailed when firebombs were tossed through their shop windows. At the same time, the offices of the Riverdale Press in New York, an outspoken local weekly newspaper, were heavily damaged by a firebomb the week after the paper ran an editorial supporting Rushdie. Even Waldenbrooks pulled the book off the shelves and executives at Viking Penguin in New York went to work wearing bulletproof vests.

Alarmed by the Islamic public's hostility to Rushdie's book, I inked out the silver title of our copy with a dark blue magic marker the day people protested on the streets of Dhaka. That day, after the Friday prayers, thousands of Muslims and several religious and political organizations observed the call for *hartal*. Carrying banners, they demanded a worldwide ban on the novel and shouted, "Death to Rushdie!" and "Burn all copies!" Hearing the demonstrators clearly berating those errant from Islam from Cole's office on the second floor of the building UNICEF occupied, I covered our copy of *The Satanic Verses* with plain brown paper before I stuffed the book between newspapers and carried it to the car as though it was a time bomb. The book stayed hidden behind papers in the filing cabinet at our house until we left Bangladesh.

Another storm gathered velocity in the Moslem world of Bangladesh in 1991. Iraq had invaded Kuwait, and the world waited for war during the season when the air turned cool.

"Long live Iraq. Long live Saddam Hussein, the greatest Muslim leader. Death of Bush. Death of America and its Running Dogs," declared the Groups of Peace of Bangladesh in a bulletin Cole brought home from the UNICEF office in 1991.

Alert to the threat of terrorist reaction, I watchfully drove past DIT 2 to where I volunteered at the Kumudini craft shop. Seeing no cause for concern, I parked the car in the compound and complimented the staff on the exquisite pottery they were unloading before I checked the display of shimmering silk saris that had just arrived. Minutes later, Selma, who worked in the shop, clutched my arm and drew me to the window to watch angry men at the Embassy of Iraq across the road raise their arms and shout obscenities against the American president. Advising me to leave the premises, Selma held the gate open as I backed the car onto the street and drove home unharmed, having avoided the stones that had badly injured other foreigners in their vehicles.

Removing the "DODGE" brass plate that covered "ISMAILI HOUSE" from the pillar of the entrance to our residence, I locked the gate and ran into the house to answer the telephone I hoped was Cole. But on the line, Joya Poti from the Kumudini craft shop warned me to "drape a Bangladesh flag over your gate if you have one, and please, please remain in your house."

Deeply concerned, I glanced from the window to see a large procession of white-robed men waving red flags advance down the street carrying giant posters of Saddam Hussein. Heart pounding, I ran down the stairs to call our little dog indoors, just as I heard shouts and the sound of rocks being thrown at the Australian Embassy. Chanting in unison, men leapt over the low wall into the neighboring compound and threw stones at the windows of the embassy. Convinced they would enter our premises, I hid in the worker's latrine as the mob threw rocks on the driveway, smashed our lights and beat the gate with sticks until they surged down the street to vent their fury on other residences and the American Club.

Responding to the extensive damage the protestors had caused, the government sealed off the residential area of Gulshan and placed protection on twenty-four-hour guard duty in tents across the street from where we lived.

Curious to see what, if any, destruction the angry crowd had caused to other parts of the city, Cole rode his bicycle to where UNICEF was located and called me from his office to say the streets were uncommonly deserted and that his ride had been uneventful. However, bulletins continued to warn foreigners to maintain alert precautionary measures from potentially serious incidents. And as MiGs flew overhead and the telephone rang with periodic warden checks, we continued our evening walks around the small pond near our home to share information with those we met.

Chapter VI
Jhorna

LATE IN NOVEMBER 1975, Cole and I decided to take a break from the troubled situation that was Bangladesh, and flew to India. The little cog -train we boarded at the New Jalpaiguri railway station chugged up winding *ghats* to Darjeeling, past hillsides covered with tea. It was evening by the time we coaxed our two tired little ones up the steps of the slope to the Windermere Hotel. Chill winds blew rain over the mountain town that evening, as we relaxed in the comfortable lounge near a blazing fire surrounded by shelves of books. At dawn the next day, soft light fused pink mist into what seemed an illusion, but soon became a breathtaking view of timeless snow that capped the lofty Kanchenjunga mountain peaks.

Later that day, the clatter of hooves on the cobblestones outside the hotel attracted the attention of our three-year-old daughter, who was keen to ride one of the old mares. Soon afterward, Cole strolled beside her as she bounced to the gait of the horse. Meanwhile, Sanjeeb held my hand tightly as we went for a walk up a steep path that led to the top of a hill, where we saw a sign that read *Shishu Bhavan*. "Let's go see the children," I suggested, as we approached the gate that led to Mother Teresa's Sisters of Charity, where a cheerful Indian woman wearing a familiar white sari with blue trim held a baby girl with huge lively eyes. "Hello, Sister," I said, holding out my arms for the child.

Sanjeeb hurried beside me as fast as his three-year-old legs could go, back to the hotel, where we found Cole and Jessica enjoying tea in the mountain sun. Excitedly, I told them God had led us to a very special child, and after lunch, the four of us retraced Sanjeeb's and my footsteps to the *Bhavan*. The sisters were eager to talk, and after some time, they acknowledged "evidence of our sincerity" to adopt the little girl I held in my arms. Strapping her to my back with a long cloth, the sisters told us to proceed to the police station to fill out the information required to travel from India to Bangladesh. This time, Jessica and Sanjeeb did not complain about the steep walk back up the hill to the hotel. Delighted with their new sister as they watched her splash in the bathtub, they raced to tell the news to the hotel proprietor's wife, who gave us a Tibetan boot as a gift and an old sheet to tear up to use for diapers. That crystal night, the constellations poured dippers of stars on our satisfaction as we sounded out names, eventually deciding on Jhorna (waterfall), followed by Amrita (nectar).

One week later, all three of our little ones, who were all under the age of three, were unwell when we returned to the Indian plains. The Indian women at the airport were sympathetic with the condition the children were in and helped me wash them before we boarded the bumpy flight for Dum Dum.

"Surely it's one of Calcutta's sultriest nights," I complained to Cole, who was equally distraught at the condition the children were in and the unpleasant circumstances at the airport. The workers were on strike. Lifting red flags on long bamboo poles, the demonstrators shouted in unison and walked round and round the conveyor belt. Weary, and having waited for hours for our luggage to arrive, we carried our bags and tired children in the direction the throng was moving and found a taxi driver, who threw some of our suitcases in the boot and tied the rest to the roof rack with rope. Jessica, Sanjeeb and Jhorna sat cramped between us as we drove through the noisy mayhem of the "City of Joy"; past piles of decaying rubbish cluttering the drains and broken footpaths choked with pedestrians. Huge billboards advertised Bollywood's latest films near makeshift dwellings inhabited by families who had come to find work in the city Mahatma Gandhi had referred to as "a violent, repressive, monstrous machine; metropolitan congestion that crushed human souls and destroyed all the good and beautiful things of nature."

That night, I was close to the limit of endurance. To add to our discomfort, Calcutta's excessive congestion seemed, as Geoffrey Moorhouse describes, "perilously close to a breakdown, collapse or whatever we choose to call the ultimate metropolitan nightmare." Stalled behind buses, lorries, scooters and taxis emitting clouds of noxious fumes, we waited impatiently for the traffic to move on until we located accommodation not far from the city center. A group of shrouded people were huddled on the sidewalk near the entrance of the hotel. Rousing the sleepy desk clerk, we checked in and made our way down a musty hall to our room. While the children slept fitfully and feverishly in a shared bed, Cole and I spoke of our concern regarding legal adoption documentation necessary to travel to Bangladesh . . . papers that had yet to be completed.

Finally we slept, waking to an overcast dawn, convinced our best course of action was to visit Mother Teresa. Cole found her that day among her colleagues at 54 A Lower Circular Road. At once sympathetic and practical, the woman Malcolm Muggeridge described as "love in action" and a champion for the dying and those less fortunate met Cole in the courtyard to hear our story. She listened attentively to what he had to say, then voiced her concern. "The matter is difficult. Jhorna was born in Kalimpong. You cannot legally have guardianship for her in India," she said. "But be aware I know that time is short and decisions must be made." Mother Teresa then led Cole down the stairs and entered Jhorna Amrita's name in the Missionary of Charity's "New Arrivals" register, adding "unknown" under the name for mother.

More confident after speaking with Mother Teresa, Cole purchased plane tickets for two to South India, where my brother John and his wife Carol taught school in Kodaikanal. It was our prayer that they would care for our new little daughter until we completed the documentation necessary to adopt our little girl. The next morning before daybreak, Cole hugged Jhorna and me close before he placed the suitcase in the boot of the taxi. Tears streamed down my cheeks as I said goodbye to Jessica and Sanjeeb who, still unwell, were about to board a plane with their father for our home in Dhaka.

"Soon, I'll see you soon . . ." I shouted as the taxi driver closed the boot and drove past women sweeping refuse to the sides of the street.

While I flew with Jhorna above clouds south to Coimbatore, sober officers confiscated Cole's passport on his arrival with the children in Dhaka. "You are in disfavor," the men in uniform said crustily, before marching him to the security desk, where he was advised to report directly to the president's office. Apprehensive about the unexplained developments, Cole hurried the children home to the watchful care of our *ayah,* Theresa, before he presented himself to President Khandaker Moshtaque, who nodded for him to sit beside him in a wide circle of foreign agency directors sitting stiffly on their chairs.

Shaking his finger at Cole as if to chastise a disobedient child, the president fumed, "You have done a wrong thing." He unleashed a torrent of angry words at Cole and told his captive audience that Cole had reported to the BBC alarming results concerning the nutritional survey he had commissioned to be done on the suffering people who inhabited the *bustee* camps surrounding Dhaka.

Following further reprimands and a stern warning, the president (who remained in power for only eighty-three days) allowed Cole to depart with the others. Relieved but shaken, Cole left the government office to continue his work in a stress-filled environment during our remaining days in Bangladesh.

Not long after Cole met with Mother Teresa in Calcutta, we were fortunate to have her visit us on our sunlit porch in Dhaka. Deep creases lined her face as the revered Messenger of Hope held my hands in hers. Known to fiercely defend her beliefs, she responded quickly to my question, "How can we possibly accomplish what you have?"

"Do as well as you can in the position and with the talents God has given you, my child," she replied, patting my arm.

Years later, the Noble Peace Prize laureate, who had made lifetime vows in 1937 of poverty, chastity and obedience, visited our home in Ethiopia and, still later, Jhorna and I paid special attention to her forceful words at a gathering she addressed in Khartoum, Sudan. Standing by until the crowds dispersed, we greeted Mother Teresa and, after a fleeting look of recognition, she placed a gnarled hand on Jhorna's head and offered her God's blessings. "Certainly we have met before," she reflected. (She had not forgotten that she had falsified the birth registry of the Calcutta orphanage in order to expedite Jhorna's adoption procedures.)

Then, one evening in 1991, when monsoon rains fell heavily from October skies, Cole and I again visited Mother Teresa, who welcomed us warmly in the slums of Calcutta. Clasping my hands tightly in hers, she spoke in a strong, quiet voice of the power and comfort of prayer, *for it is by forgetting self that one finds oneself.* About issues close to her heart. About the scourge of Aids, those dying in the gutters, and unborn aborted babies.

Mother Teresa left me with unanswered questions that day, such as why there was the incessant clash of civilizations and relentless poverty, and the causes and devastation of war—questions she told me we would never answer, but that my search for truth would lead me on my own journey to do what God had given me talents to do.

In 1976, Oxfam asked Cole to become the Field Director in Ethiopia, and the time had come to leave our home in Dhaka. Chattering children followed Jessica, Sanjeeb and me up the grassy slope to Dhanmondi Lake, where we had gone to say goodbye at the location of many walks. Back at our house, we pushed open the gate to find Theresa and Veronica holding bunches of lavender and white cosmos blooming along the drive. Pressing the flowers into our hands, the women wept as Jhorna toddled sympathetically from one to the other to wipe away their tears, for they had been part of our family for two years.

Chapter VII
The Lion of Judah

AFTER HAVING VISITED Ethiopia as the acting director of Oxfam America, Cole had described the country's rich history and proud, handsome people to me, and I was keen to see the country for myself.

Circling over the ruggedly beautiful highlands, the pilot swept the plane down the runway, past women slashing grass with scythes. In the distance, the Entoto Hills were silhouetted darkly against the sky behind Addis Ababa. Clattering down the stairs ahead of Cole carrying Jhorna, Sanjeeb waited on the tarmac for Jessica and me, then raced ahead to the airport, where women wearing traditional white cotton *shemmas* with colorful decorative borders smiled and reached out for him. Welcoming us with a bow and, "*Tena yestilign, endemin allu,*" Oxfam's driver, Ato Belai, helped us complete the necessary paperwork before we loaded ourselves and the luggage into the Land Rover. The road to town was lined with *bahr zaf,* tall stately eucalyptus trees that had been introduced to Ethiopia around the turn of the century when the requirement for firewood and building materials far surpassed what was locally available.

Much of the Ethiopian highlands had been severely deforested, causing rich topsoil to wash into riverbeds. Too poor to afford fertilizers, farmers continued to plow their fields, but when the rains failed again in the early 1970s, starving people slaughtered their livestock in order to survive. Tens of thousands perished during the drought that affected Wallo province in the north of the country as day after dry day turned the Ethiopian land to dust. Still, the Ethiopian government under Emperor Haile Selassie made no request for international aid. This started the process for sections of the army, members of the educated elite and a number of junior officers to form the Dergue, led by Lt.

Colonel Mengistu Haile-Mariam. Mengistu, who was from the Galla tribe, wore inch-high platform shoes to add height to his five foot three inches, and vowed that those opposed to the Marxist/Leninist revolution would have their blood spilled. And they did.

With a ruthless taste for power, Mengistu shot dead his nearest rivals to avenge the blood of his comrades in the ongoing insurgency. Thousands of counter-revolutionaries were killed. Thousands of others, including Haile Selassie's relatives or those with ties to the royal family, were jailed.

Eighteen years after Haile Selassie was overthrown, Uganda's daily *New Vision* reported that the Lion of Judah's remains were exhumed from a secret grave under the floor of the toilet in the grounds of the Imperial Palace. The article went on to say that Eshetu Tekle-Mariam, Selassie's pageboy, who was allowed to sleep with the Emperor during his detention, had said his master did not complain of any ailment before he was taken to a nearby guardhouse where he was ordered to sleep.

"At that moment," Eshetu was recorded as saying, "the Emperor must have guessed he was near death and he turned his face towards a wall, facing the same direction as a nearby church before he made the sign of the cross. He wept then and said 'I have done everything I could for my country, let history judge me.'"

The following morning, Eshetu found the emperor dead on his bed. The government claimed the emperor had died of natural causes but Mo Anbessa from the pro-monarchist movement claimed Haile Selassie had been suffocated with a pillow soaked in ether.

A Taste of the Dergue

Tim Briely, Cole's predecessor, kindly offered us his apartment in the IRESCO building that overlooked the railway station the French had built in 1929, until we could find our own home. After tucking the children into bed, I looked forward to having time to look through the books Tim had on his shelf and declined Cole's suggestion to go for a walk. Engrossed in reading about Ethiopia's rich culture and recent history, I heard the door slam. Out of breath, Cole threw himself on to the couch next to me.

"It's a long sprint up six flights of stairs," he panted, grinning at me. "Near the Ras Hotel, men tried to wrestle me to the ground, I suppose to pick my pockets. But I threw them off."

Although we had been warned it would not be easy to find a place in which to live—for not only had the government nationalized financial and industrial institutions (without compensation), but neighborhood associations had been put into effect and *kebelles*, or worker peasant defense units, had taken over private properties—Cole found us a house near the old airport that overlooked a steep ravine. Soon after, in need of purchasing household effects, the children and I made our way through the sprawling *mercato* to barter with the merchants there. At the bottom of a muddy hill, a frail old shop owner opened the shutters of his small shop to let in sunlight that exposed dozens of intricately woven baskets piled high on the earthen floor. Offering the children leather-covered stools on which to sit, the old man helped me find a basket for the purpose of holding *injera,* local bread. Satisfied with the price I offered, the shopkeeper presented the children with an animal carving that he wrapped in paper along with other handcrafts.

Always the cautious one, Sanjeeb kept an eye on what we had purchased while the girls and I selected hand-woven fabric and two large clay vessels, which the shop owner then helped carry to the car. Puzzled, but pleased with our choice of "only local things," Ato Belai packed the artifacts into the boot of the Land Rover before he drove us down the road to visit the Institute of the Blind, where, he said, carpets were woven with wool. Fascinated by the skill of the sightless men who worked their looms with seeming ease, the children found excuses to delay our departure. But when the light began to fade in the dusty room, I convinced them to follow me to the car with the promise to purchase a rug with an *ambosa* woven into it—a lion in black and brown.

At the top of the hill, Ato Belai had started coasting down the broad boulevard when, with disbelief, we watched two tanks and other military vehicles advance threateningly toward us on the wrong side of the road. Traffic stopped abruptly as shots rang out. Screaming, pedestrians fled in all directions, and I ordered the two older children to crouch down in the back seat and pushed Jhorna with me to the floor. Peering over the dashboard, I watched a terrible drama unfold. Two persons who leapt from a car that swung to the curb were brought down by bullets fired from one of the military vehicles. While blood spread darkly on the sidewalk and more shots exploded around us, I closed my eyes and pled with Belai to do what he had to in order to get us out of the mayhem. Belai managed to turn the wheels over the curb and drive up the sidewalk to a side lane. Huddled on the floor of the Land Rover, the children studied my weak grin with apprehension, but soon bounced back on their seats in their usual manner and, as time released tension, Ato Belai pondered aloud the damage caused to his vehicle and our emotions.

"Political ideology, whether it is imported from east or west, does not fill stomachs," David Lamb states in *The Africans.* "It merely usurps energies that should be devoted to national developments. It leaves a numbed people mumbling foolish slogans at the command of their presidents, and if that ideology is Marxist, it removes all individual economic incentives."

Breaking the feudal land tenure system, the Marxist government allowed the Ethiopian peasants to own the land they cultivated, with the result that food production rose substantially for the first year or two following the revolution. But then incentive and productivity sharply declined, due to, among other things, mounting taxes to pay for the entrenched civil war in the north, political repression and censorship. Unhappy with the trend, discontented bands of armed men roamed the streets to fight their own battles with those who opposed their views.

One moonlit night, disturbed that our Alsatian barked and barked, I pushed back the curtains of our bedroom to watch dark figures move stealthily down the gravel path behind the house. Thereafter, we locked our gate.

Increasingly aware that small aid agencies and missionaries had lightened their personnel presence in response to the country's deteriorating security situation, I hurried the children down the back flight of stairs one late afternoon and left Cole to confront the armed Red Terror in their futile attempt to find guns on our premises. The Economic

Commission for Africa official who occupied the house across the wall from us, however, was not so lucky. Severely reprimanded for having the weapons he illegally owned, he lost his guns to the *kebelle*, who issued him a receipt and departed with a warning.

Despite the international community's increasingly distrustful opinion of the Dergue, Oxfam enjoyed an amicable working relationship with the civil service, peasant's associations, Relief and Rehabilitation Commission and others who not only had the power but the mandate to control education, health and security as well as the collection of taxes. Together with Simeret, who worked with the government horticultural division of the agriculture department, Oxfam encouraged the *kebelles* to cultivate small community vegetable plots with a project that flourished on urban Addis land. Offering technical assistance, John Collett and Steve Bonnist helped set up a water development unit with an Oxfam support group that traveled from mission to mission advising the locals on how to select a site, dig a well by hand (in great demand in a country where clean water was so scarce) and line and protect it. One demonstration well was dug in each of the localities, after which the project extension was left in the hands of the local mission or organization.

Acting as vice-chairperson of the Christian Relief and Development Agency, Cole met regularly with, among others, Fathers Kevin Dahoney and Angus and Jack Finucan (whom he had previously worked with in Nigeria and Bangladesh), who ran the Irish NGO, Concern, to coordinate relief and development work for the various voluntary agencies and to receive bilateral relief allocations that would otherwise have stayed in the hands of the government.

On a day bright with sunshine, we drove to the hills that turned golden with *adey abeba* flowers during the Ethiopian spring. Spreading our blankets on the rise to survey the expanse below, we greeted the women who labored up the hill with bundles of wood strapped to their backs. The women, who had expressions of dignity on their weathered faces despite their obvious poverty, urged their children to accept the bread we offered. The children, who had flies clustered around their eyes and lips, hesitantly came forward to take what we offered. One young girl, who had not passed more than sixteen *meskal* seasons, had her hair plaited into intricate braids. The silver cross she wore around her neck caught the rays of the sun. The icon she wore represented the Christian faith that had been brought to the highlands of Ethiopia by the Egyptian Copts. When the religion of Islam began to influence the people, the Abyssinians withdrew, and for centuries isolated themselves from the rest of the world.

When the shadows lengthened, we folded our blankets to journey down the hills past jubilant crowds dancing along the road waving branches. At home, we found that Malaku had stacked wood for our own celebration. Delighted with the idea of festivity, the children helped him weave flowers they collected in the hills into the image of a cross he placed on the top of the sticks. That evening, we lit the fire to commemorate the finding of the true cross and flames leapt into the air with our merrymaking. Seated on the ground opposite me, Zenovitch's beautiful features glowed in the flickering light. And when the blaze died down, she invited us to her small room, filled with smoke for the preparation of *buna,* a custom she was well acquainted with.

"*Yikemetu,*" Zenovitch said, offering us stools before she selected the coffee beans she roasted on a flat iron pan. Throwing gum resin on the charcoal, she then pounded the beans in a mortar and pestle and carefully poured the powder into a clay vessel. Adding water, she placed the pot on the glowing charcoal to steep.

After talk and a little time, the delicious aroma of coffee blended with the sweet, heavy fragrance of the resin. Pouring the strong brew into tiny clay cups, Zenovitch told the children, "*Tinesh,*" only a little for you. Pleased to take part in the occasion, the children sipped the coffee slowly and licked the generous spoonful of sugar that settled to the bottom.

Coffee is indigenous to Ethiopia and still grows in the Kaffa region," Cole reminded me, as we watched our little ones enjoy the experience more than the drink. But it was not until the thirteenth century that coffee beans were brewed into what we were now drinking.

* * *

"*And, hulet, sost, arat, amist* . . ." I heard the children count one afternoon as they laid twigs in the sand near Weaver John and Wet Top, our rabbits, who hopped about casting pink eyes over the proceedings. Pushing the last tomato plant into the soft soil, I walked around the house to remind Malaku that it was time to place Jessica, Jhorna and Sanjeeb, in that order, on Kokeb's strong back and lead our horse down the ravine, through the stream and back home around the eucalyptus trees, as was our custom when the weather was fine. And when the sun slipped behind the hills, I ran water into the tub and watched the children splash before they pulled on their "zoot" suits with slippers attached, "that kept out the poochis," as Sanjeeb conjectured. Then, despite hosting frequent houseguests, Cole and I spent evenings with the children with stories and song near a warm fire, seated on the lion rug that the blind men had woven.

Other days, when Cole found the time, we drove from the chill mountain air to where frangipanis dropped fragrant flowers and bougainvillea spread brilliant colors over low walls. Donkey carts loaded with hay and produce plodded slowly ahead of us down the dusty hot streets of Debra Zeit before we followed dirt roads that led to the lovely volcanic Lake Hora. There, we walked the trails, followed by little children who brought us bunches of wildflowers they exchanged for a coin. Advised not to swim in the inviting water, as snails harbored schistosomiasis, which attacked the liver and intestines with results that took time and medicine to heal, we skipped rocks across the gentle waves that lapped the shore.

Turbulence

Little market towns in the arid region of the Ogaden in Ethiopia were divided into the Somali quarters, where the currency used was shillings, and the Amhara quarters inhabited by police, army and a few highland traders who used Ethiopian *birr* as their currency.

Tribal differences simmered until 1977, when the Somalis, who were defiant and thrived on raiding even in normal times, invaded the Ogaden—an area ethnically their own.

Scorching winds parched the Ogaden as Cole drove the last miles to visit the Oxfam project Rob Reese had undertaken to introduce spread irrigation and inexpensive rudimentary agriculture to the pastoralists there. That evening, while they sat near a flickering fire, a lean, barefoot Somali emerged from the shadows to warn Rob about military men infiltrating the area and sure signs of an approaching battle.

Long after the informer left, Rob and Cole sat near the warm coals and discussed the matter, which certainly held risks, for it had only been a matter of weeks since Cole had driven north, in what he described as a "bone-rattling ordeal," to Selikelka, to evacuate Oxfam's water engineer, Duncan Alsebrook.

But apprehensions faded in the new morning light, and Cole filled the vehicle with fuel in Kelafo to commence his two-day return journey back to Addis.

Navigating on tracks left by the tanks military men in Gode had warned him about, Cole contacted the military unit at the town of Kebra Dahar, who confirmed there had been recent fighting to the east of the north-south road.

"Proceeding is dangerous," they told him. "But if you go on, there is an eleven o'clock curfew in Jijiga, so arrive before then."

Keeping an anxious eye on the indicator that showed the fuel tank would be very close to empty by the time he reached Dega Haboor, Cole sped north over the corrugated road, past armored combat vehicles with mounted cannons that gave evidence of a recent battle. Well after curfew, not far from the outskirts of town, the fuel tank registered EMPTY, and Cole coasted down the long slope of a hill that led to the checkpoint, not bothered by the order to halt.

Heaving a sigh of relief, he asked the suspicious guards the whereabouts of a lodge to lay his weary head. Locking the Land Rover, Cole walked the rest of the way to town, where he found a vacant room and fell into deep sleep, until he was wakened by the thud of artillery coming from the direction he had traveled.

Filling the Land Rover and spare jerry cans with diesel he obtained from army men, he proceeded to Harar and Dira Dawa and arrived home before curfew. That evening, seated around a warm fire, Cole related tales of military tanks and dusty roads to the children and me, until we were interrupted by staccato interference over the radio, and Rob's voice saying, "I need to get out."

"The Ogaden has been invaded from the east and a battle is almost certain to follow," Cole said. Dreading to hear what I already knew, I heard him say he would need to charter a plane to evacuate Rob.

Before sunrise the next morning, Cole drove to Bole airport to find Don McClure, his wife and Don junior seated in the Air Mula plane he had chartered. Inquiring of their intent, he was told they wished to fly with him in order to drive their vehicles back to Addis.

"The road is not safe," Cole warned the missionaries, telling them of his own recent experience. The men were not convinced, but further persuasion influenced Mrs. McClure not to travel.

Silent with their own thoughts, the men bounced through the turbulent air in the light aircraft. Circling the plane before he landed, the captain halted near to where Rob Reese stood together with his Quaker friend on the gravel airstrip.

"The Somalis have already infiltrated the area," Rob cautioned. "It is said there will be big trouble tonight."

Tilting his head towards the dark skies, Captain von Rosen recommended they wait until the following morning to return to Addis. But Rob convinced them otherwise, and they decided to proceed back at once. All of them, that is, except the McClures, who made the decision to stay behind.

"We have worked with these people for so many years," they said confidently. "We will travel back to Addis by road."

"Your choice," the captain retorted shortly, as he eased himself into the cockpit. Lifting the plane around dark cloud towers, von Rosen flew the plane through heavy skies. Seated next to him, Cole listened to the stories the captain had to tell of how he had flown medicine and food to the guerrillas who held out against the Italian troops in western Ethiopia. Grateful for his assistance, Emperor Haile Selassie had honored von Rosen by presenting him with a coffee plantation that the pilot enjoyed for a few months every year. Later, the emperor asked the count to organize the Ethiopian air force; a task that von Rosen carried out from 1946 to 1956. Years later, word reached us that the captain's career had abruptly ended when guerrillas shot him dead.

It was early dawn when a strong knock on our front door roused us from our beds. Pulling on a wrap, I followed Cole down the stairs to hear someone say, "Don McClure was killed last night."

Stunned, we listened to the rest of the story.

"Somali men crept through the night and surrounded the mission compound," the messenger said. But before the armed men entered the house, they forced the missionaries to the garden at gunpoint and shot their captives at close range. Two of the three men died on the spot. Only Don junior lived to tell of the ordeal.

"It was as though the bullets tore through me," he later recalled. Unconscious for some time, Don awoke from his near-death experience and crept on his hands and knees towards the Wabi Shebele River, where he buried himself in the sand until he was found at daybreak. The women at the mission survived their own ordeal and were evacuated in a DC3 in a state of shock and anguish.

Where Lucy Was Found

Not far from where Cole and his colleague David Breyer had driven to visit Oxfam projects in the region of the forbidding Danakil Depression (the lowest point on the earth's surface, which dips to more than 328 feet below sea level), they traveled on to an archeological site in Hadar where the earliest known hominid (at the time), known as Lucy, had been found in 1974. Dozens of tents offered the relief of shade, where the men stopped to talk to the French and Americans excavating the site, but, like the inhospitable terrain they had driven through, those exhuming remains were hostile with their reception until further request provided the weary travelers with a tent some fifty meters from the main camp. Inquiring as to why the hominid remains were referred to as Lucy, Cole was told the Beatles' song "Lucy in the Sky with Diamonds" frequently blared from the camp tape recorder during the dig and the name seemed suitable. Also known as *Dinkenesh*, "thou art wonderful," Lucy had lived some three million years ago, when the Afar triangle was a well-watered region of forest, grassland, rivers and lakes.

That night the most unusual occurrence of rain washed the expanse of scorched landscape as far as the eye could see. The *wadis* or riverbeds were now impassable and Cole and David were told they would have to wait in order to drive the normal track to the main road. Concerned about his flight to the Yemen the following day, David Breyer, the Oxfam secretary for Africa (later to become the Director General of Oxfam), requested that a guide accompany him to walk the seventeen kilometers to the main road along the Awash River, which provides the only water for the Danakil's nomadic Afar herdsmen, who, as Colin Willock writes in *Africa's Rift Valley,* believe that "he who controls the water controls the Danakil. Any intruders who might threaten their lonely tenure are murdered and castrated, not necessarily in that order."

By noon the same day, soaring temperatures baked the soil, allowing Cole to follow the much longer route that led to the tarmac road connecting the port of Asab to Addis. Thanking the Afar tribesman who directed him, Cole watched the tall, proud figure become part of the mirage that shimmered towards the horizon and continued toward the bridge where he hoped to find David.

A single camel sauntered down the road near the army checkpoint. Inquiring of the military man whether he had seen a white man walking from the west, Cole was told, "He has been arrested. The man can be found at the officer's post—that uniport there on the other side of our compound." The man pointed his rifle to the opposite side of the bridge.

Fearing hours of interrogation and an unfortunate outcome, Cole hurried toward the military post and stopped the Land Rover. Straightening his shirt, he marched into the commander's office, where he found a distressed, disheveled David standing before an officer seated behind a large table. Greeting the official, who was in the process of rising from his chair, Cole slipped his identity card onto the desk and swooped up the documents spread on the table. With a nod and a "thank you" to the officer, he told David, "Let's go," and the two of them disappeared into the white sunlight in the direction of Addis.

Totally exhausted, the men pushed their luggage through the front door of our residence just before the city's ten o'clock curfew.

With Hope for Our Return

Fields of indigenous grain, *teff*, used for making Ethiopian bread, grew at the side of the road we followed. Women beckoned us to buy the zucchinis, tomatoes and eggplants they had grown. Stopping the car for me to purchase a basketful of fresh produce, Cole placed the vegetables on the back seat between our young ones, before he urged me to discuss the option he presented—either to move to Nairobi or take a one-year sabbatical (something Cole intended to do in a few years anyway).

"Perhaps the Director of Overseas Aid in Oxford listened to David Breyer and decided Ethiopia was no place for us as Americans," Cole surmised, as I watched the wind chase the grain alongside the road. Changing the subject to keep from having to make a decision Mengistu Haile Mariam's Red Terror had influenced, I was aware of the country's steadily deteriorating security situation. Recently, Oxfam's personnel had been forced to leave their locations in Makelle in Tigre, the Ogaden and eventually the entire Afar and Isa areas of the Danakil.

But Cole pursued the topic, and handed me the envelope he had kept in his pocket for some days. Postmarked Seattle, Washington, it was from the university there, saying the School of Public Health had accepted Cole's application to further his graduate courses. Anticipating my reaction, Cole drove down the gravel road to our house and led me to the back stairs near the kitchen door, where we leaned our heads against the cold support of the concrete wall and watched our children play in the afternoon sun.

"We'll store our furniture in the event of our hopeful return," Cole said, without conviction.

I heard his words, but not with my heart, as my fingers traced the pattern of the worn tile steps.

So we took our pictures from another wall. Malaku rode our horse to his father's village. "Where the grass is much greener," he told the sorrowful children. Mickey followed us with great sad Alsatian eyes. Stroking his head, I told him he would bark with new friends at the doctor's residence where he would live. Carrying the last suitcase to the car, I walked back to the house to take a long look at the roses that spilled red clusters over the wall. "We will meet again," I heard Zenovitch say as tears flowed down our cheeks. Embracing the children on her knees, she bid us "*Dehna hunu*," and closed the gate behind us for the last time.

We never did return to Ethiopia. After Cole completed his graduate studies in Public Health from the University of Washington, Oxfam requested Cole to cover Tamilnadu, Kerala and Sri Lanka from Madras (now known as Chennai).

Chapter VIII
Ancient Rhythms

KNOWN AS A CONSERVATIVE bastion of social and religious concerns, Madras surrounded us with familiar sights, sounds and smells from my childhood days and what Cole and I had experienced together. Oxfam's driver, Xavier, drove us slowly through the urban congestion, as we called our curious children's attention to what we remembered. Crowds walked past stores selling typewriters, bicycles, refrigerators and ceiling fans manufactured in their own country. But for a businessman to build a new factory or expand an old one was not an easy task, for he had to request an official for a license that often took years to obtain. And worse, as stated in the *Asia Week*, "the system promoted corruption. Rich men became that way by seeking favours."

The driver, who had been sprawled out in the front seat of his taxi, sat upright and moved his car when Xavier honked at the entrance of the Hotel Dasaprakash. Summoning porters, some of whom recognized us from years gone by, we exchanged greetings in Tamil and handed them our luggage to take to reception. Scampering ahead of us, the children ran up the stairs past garlanded pictures of Hindu gods and goddesses to the suite on the third floor that would be our temporary lodging until we found an exceptional dwelling in Poes Garden—just a few houses down from J. Jayalalitha, who later became the Chief Minister of Tamilnadu. This was the woman Christopher Thomas, in the *Times* of 7 September 1995, described as the Empress of Southern India. Surely she was an empress as far as lavish ceremonies were concerned. For she hosted an extraordinary wedding for the nephew of the man she was living with (officially described as manager of the Chief Minister's household) under floodlights that blazed over seven miles. The extravagant occasion, at which hundreds of sheep and thousands of chickens were slaughtered, was so vast, Thomas reports, that, "parts of the power -starved city of Madras had to be blacked out." New roads were paved for the wedding set, old ones widened, trees felled, new telephone cables laid and police check posts set up to keep out the poor majority. The excessive celebration, which cost at least fifteen million pounds, was attended by some two hundred thousand people, despite the local law that prohibited attendance of more than 250 guests at occasions. The law was designed to stop vulgar displays of wealth. But wealth was relative.

Certainly the contrast of what we enjoyed in the up-market residential area kept us significantly apart from those who survived in the overcrowded slums along the Cooum River that poured raw sewage through the fourth-largest of Indian cities. Madras teemed with people, and instead of the wide-open view we had enjoyed in Africa, we now looked on the walls and closed window drapes of our neighbors. But in time, the oleanders and jasmine flowers I planted in pots scented the verandah near the tipu tree that dropped yellow blossoms.

Shortly after I spread brightly colored cushions on the living room floor, smartly dressed visitors requested the use of our premises for scenes in a movie they were producing. Although we toyed with the idea of hosting an Indian heroine dressed in sequined *saris* dancing on our marble floors, we declined the film director's offer and suggested they find another venue for their stars. The producers, however, were not easily dissuaded, and asked to use the 1946 Jaguar we had purchased from a friend for the hero of their film. Again we rejected their offer and drove our "car of utmost class," with leather seats and a hatch that opened, to and from Harrington House School to pick up the children from classes. Until one day, the front wheel of the car rolled to the side of the street, after which the Jaguar became known to us as the "car of utmost crash."

On occasion we drove the Jaguar through the snarl of auto-rickshaws, bullock carts, buses, lorries, pedestrians, cars and wandering cattle past the artist village, Cholamandalam. Farther down the coastal road, the children played in the Mahabalipuram sand near casuarina trees that swayed in the sea breezes, and clambered on the sculpted bull that was said to be Shiva's *Nandi.* When it was time to drive home, Cole opened the Jaguar's hatch and the children stood up to catch the salty air from the restless sea. It was an unpopular decision for all of us to sell our classic car when we reached the end of our stay in Madras.

"It must be the city's pollution that contributes to your children's respiratory disorders," the kindly pediatrician told me one day on one of my frequent visits to consult her about one or the other of the children's bronchial complaints. I sighed, and told her that only the day before Jhorna had observed the filthy water in the pail Antoniamma used to mop the floors of our home.

"It has to be the traffic emissions that hang in the humid air," the doctor continued, as she wrote down the prescription.

Convinced Jessica's lungs would collapse from coughing one later evening while traveling on a train, I waved unhappily to Cole, who continued his journey to Kerala on another train, and turned my attention to our three ailing children, all of whom were flushed with fever. No sooner had I cleaned up the mess from one child's vomiting than the next one would cry and do the same. Tight with fatigue and concern, I sponged their hot bodies with wet washcloths until we reached Central Station, where I requested two porters help me carry the sick children to find a cab. This was not easy, as Madras was flooded from recent rains. Throngs of drenched people holding black umbrellas splashed past the taxi, causing the driver to shake his head sympathetically as he drove us to the comfort of our home. The children recovered after a day or two, and were eager to return to Harrington House School where the British headmistress carried on classes despite the water that covered the premises. A practical woman, she told the parents, "Doesn't this happen every monsoon season?" And she allowed the children to float the boats they folded from paper in the muddy water, probably contaminated with the city's sewage that overflowed from open drains.

Impatient with systems crippled by corrupt management and sporadic results, on occasion the city's residents took to the streets to demonstrate their concerns. One afternoon, after tea, the children and I left Cole at the squash court and drove the Land Rover over the bridge toward the marina that bordered the sea. With dismay, I watched hundreds of young men, who had minutes before been walking on the sidewalks, link their arms and brazenly walk toward us down the middle of the street.

"Lock your doors," I told the children, realizing that it was impossible for me to turn or reverse. The rowdy youth, who were brandishing sticks, mounted the hood of the car and clambered onto the roof. Leaning over the windshield to leer at me, they gestured rudely and laughed at my confusion. Agitated beyond measure, I glanced back to see the children crouched on the floor. Having no choice but to drive forward, I stared straight ahead and was immensely relieved when one by one the unruly men jumped from the car. Able now to direct the children's attention to the blue of the Bay of Bengal, I drove toward families dressed in colorful attire promenading on the beach near fishermen who fought the waves to pull in their catamarans piled high with struggling fish.

Toward an Integrated Whole

A number of projects Cole supported through Oxfam represented idealistic change, such as the one Richard Stanley set up in the city of Bangalore called SEARCH, which provided meaningful orientation and leadership training for young men and women, who in turn set up other local organizations for social change.

One morning, while seated at the desk in the room of our residence we used as the Oxfam office, I pushed aside the typewriter I had been using to type Cole's report and heartily welcomed our friend who walked through the door.

Small of stature, Dr. A. T. Ariyaratne, the founder and director of the Sarvodaya movement in Sri Lanka, was years later presented with India's highest international peace award for humanitarian service he modeled after the teachings of Mahatma Gandhi and traditional Buddhism. With more than three million members in eight thousand villages, Ariyaratne's movement trained over forty thousand Sri Lankan youth to improve their self-reliance through community participation with health and nutrition programs integrated with spiritual and moral development.

"Now it is your turn to come and stay," Ari insisted, following his short visit with us en route to some meeting or other. Not long after, we were able to accept his offer, and flew to visit his family in Moratuwa, south of Colombo, where he and his wife vacated their bedroom to accommodate our family of five. With yet another gesture of generosity, Ari loaned us a vehicle for the children and me to accompany Cole on a tour of Oxfam projects around the island. It was pouring rain when we left the house. Driving near rice paddies that led up hills one thousand feet to Kandy, we continued east to Batticaloa, where we found accommodation in a rest house nestled in coconut plantations that stretched miles along the oceanfront. Fresh breezes greeted us at sunrise and the children darted barefoot behind the palms, not at all prepared to drive on to Dr. Sellaiah's house and his reminiscing about the cyclone that had devastated the area in 1978. But, amid visiting projects, we found time to bathe in the seawater and watch marine life in warm coves. Then, when light wind brushed the sand and the sun sank in the western sky, we coaxed our children from watching beetles roll dung into perfect, round balls to walk down the path to yet another night in a guesthouse, this time to the one at Salt Lick Farm, Nilavilli, northwest of Trincomalee.

Many years later, Ariyaratne was invited to the meeting I accompanied Cole to conducted by UNICEF in Agra to plan a strategy for children for the 1990s. Seated at the back of the large conference room, I was an observer amid talk of ratios, rates, new technology, per capita investment, relative returns and policy considerations, among other things. Keen to watch Ari's reaction to what was being discussed, I noticed he was unusually quiet, until the closure of the meeting, when he was requested to present his views. Supporting his head with one hand, he lingered over his words. "The speakers have been preoccupied with only half of mankind's problems," he said. "The physical. You have totally ignored the spiritual and the enrichment of the spirit. The strengthening of inner resources."

"Bravo, Ari," I applauded silently, absorbing his words like water on a dry sponge. Many of the problems the participants expressed, I thought, reflected what Ari so aptly stated.

Speaking on about the spate of killings during Sri Lanka's civil war and the injustice toward certain groups in the world, Ari suggested to the attentive men and women that moral values be encouraged through ethical social behavior if populations were to improve. Voicing his concerns regarding greed generated by standard developments compared to selfless service and sharing, Ari ended with:

"Ethnic and religious identities are used to describe impediments to development, but seldom if ever pursued, supported and encouraged to identify solutions to inequality."

I was sad to leave what we had come to know in Madras, and the thought now of sorting and packing our belongings—two years exactly from the month of May, when we had stepped off the plane in Madras—was a weighty one; particularly since I was still recovering from a bad bout of hepatitis.

The children, however, were excited at the prospect of a new setting. One day, as they were selecting what possessions to keep and what to give away, I heard Sanjeeb murmur as he hugged our pariah dog's russet-brown head, "Don't worry, Jasper," and my eyes misted over. In the past, I had tried to cover my sorrow from the children at partings from a cat here, a dog there, but now we spoke of finding our little dog a good home together. Vowing not to have another animal that we would have to leave behind, I knew that would never be the case, as all of us cherished and needed our pets. Sensing our impending departure as all animals do, Jasper followed our movements with a drooping tail, aware that he would no longer brave the night with our watchman, Narayan.

Narayan's "Yellow Cat" was a different story. None of us dared approach the sinewy cat with green hostile eyes that wound around our watchman's spindly legs. Guessing the cat's age to be maybe ten years, and his own at seventy, Narayan frequently stooped to stroke his devoted battle-scarred friend, while Jasper and the rest of us stood at a distance, for the cat had wicked claws and trusted no one but his master. Despite my offer of a substantial room, the two of them lived in a thatched hut that leaned against the back wall of our compound. Now, with our imminent departure, Narayan would be without his monthly salary, and the children and I talked of these concerns as we prepared to leave what we had shared at Poes Garden. But we looked forward to seeing Grandma . . .

Chapter IX
"Nabad Gelyo": May You Enter Peace

And in a random scorching flame of wind
That parches the painful throat and sears the flesh
May Allah in his compassion, let you find
The great-boughed tree that will protect and shade.
—Mohammed Abdullah Hassan, *The Prophet's Bell*

JUST A DECADE after Cole and I had been on the first civilian scheduled flight into war-ravaged Port Harcourt in former Biafra, Nigeria, and seven months before we were posted to the war-ravaged land of Uganda, the children and I again visited my mother, who, after burying her beloved husband in India, continued her work there to complete fifty-seven years of mission service. She then retired to Hillsboro, a small town in Kansas, with wide, empty streets that crisscrossed near neighbors, some of whom canned their own food and embroidered patchwork quilts. One summer night in 1980, after I had tucked the children into bed, Mother and I sat together on her wooden deck. Now and then, flying insects collided with the screen door and spun helplessly upside-down on the checkered cloth that covered the wooden table. Pushing a beetle back onto its legs, I watched it lift its wings and fly, yet again, toward the naked bulb. The telephone rang, and I went indoors to answer it.

"It was Cole," I told Mother when I came back out.

"Good news?" she asked expectantly.

"Guess so," I said, pulling my chair closer to hers, for the news meant leaving Mother. The passing years had made it no easier for me to leave her than the time I had given my parents a tearful hug before departing for boarding school.

"There's drama involved," I told her. "The Head of Human Resources at UNICEF Headquarters in New York who interviewed Cole for the job seemed annoyed that he was white, male and American, and said that they would not pay for the children's and my airfare. But it looks as though they have offered him a three-month interim contract to work for UNICEF in Somalia. That is, if he passes the CIA security check . . ."

Not prepared to take the offer, Cole informed UNICEF that if they did not pay for his family's tickets, he would accept employment with Save the Children Fund, U.K. as their Regional Director in Nairobi. That afternoon, we had tickets for Somalia and Cole had his UN passport in his hands.

And we were on our way. This time to the Horn of Africa.

Had the Somalis been able to retain their ethnic and cultural distinctions despite years of colonizers and partitioning? I wondered as I watched the emerald waters from the window of the jet before we landed on the runway just outside Mogadishu, said to be the oldest city in sub-Saharan Africa. The children were anxious to leave the aircraft, and scrambled down the stairs to the tarmac ahead of Cole and me. Immigration procedures took time, but finally we greeted Mohammed, who drove us down Mogadishu's sun-baked streets, past lean men herding camels, to the Uruba Hotel.

That evening, we dipped bread into oil and vinegar at the dining table while Mr. Bonani, the UNICEF Representative, informed Cole he would be handling supplies and logistics. "A senior staff member of mine, a woman from West Africa, has taken on the position you were assigned to. Since," he continued, directing his comment toward me, "your husband does not have a proper United Nation's contract as of yet." Shifting uncomfortably, I watched Cole swallow these words as easily as stones, for he had accepted the position of Program Coordinator.

An hour or so later, as we urged our tired children up the long flight of stairs to our hotel room, Cole justifiably fumed that he wished he had not turned down the job with Save the Children in Nairobi. "I simply am not going to handle supplies and logistics after running country programs in Nigeria, Bangladesh, India and Ethiopia," he objected. "Good grief!"

It had been a long day.

The next morning, light rain kept the children and me indoors, while Cole went to investigate his office procedures. Wind moaned through the uneven window frames and rain seeped under the door of the hotel that led to the open porch where I strung a line to hang the dripping clothing I had washed. The children and I then curled up on one of the beds and read stories until the sun came out.

Convinced it would take time to find a house to live in, a day or two later we moved to a hotel with a courtyard surrounded by age-old trees and better food. Diriye Mario hung about the entrance of the Creche del Sud and introduced himself as dealing with anything that was profitable. "I know of an Italian villa next to President Siad Barre's residence. It has been promised to a German couple," he told us, flashing a disarming smile that exposed gold-capped teeth. "But I am not deterred by obstacles." The next night, Diriye requested Cole join him in marathon negotiations with the qat-chewing Somali landlord, which resulted in us moving into our new home. It had a wide-open courtyard and a splendid view of the sea and the lighthouse occupied by our friends, and other whitewashed buildings that bordered the harbor.

The heat of the day coaxed us to spend quiet afternoons indoors, and when the sea breezes ushered in the evening, our frequent visitor, Diriye, admitted, albeit guardedly, that in fact the way he made money was by dealing in ivory and gold. But because of present laws and regulations, our friend complained, he found it increasingly more difficult to obtain these illegal items.

"Good," the children and I chorused together. And although we strongly advised him to change his pursuits, Diriye paid us no heed, for he was attracted by anything that offered him quick, lucrative deals. Jerking his head toward the big neem tree, he reminded us that we lived only a stone's throw from the president's residence.

"People in uniform and high places," he grumbled, "want the profits of such items for themselves." Diriye was referring to Siad Barre, who had seized power in a coup nine years after his country received independence in 1969. Washington backed the Somali president as an obstacle to Soviet expansion and to control the eastern access to the Red Sea, although Barre stirred up feuds among Somalia's clans and divided his opponents in order to stay in power. As long as Moscow supported neighboring Ethiopia, Somalia could count on assistance from the West. But, together with the Cold War rivalry and Barre's increasingly brutal regime, dominated by his Marehan clan, popular resentment grew, and even the poets, who traditionally memorialized Somali events in verse, called attention to the matter of potential unrest. So pleased with the panorama from our villa were we that we paid little attention to the old house with high ceilings, and used it only to sleep in. The kitchen, which was separated from the house, had an excellent view. I enjoyed the view while keeping the charcoal burning from morning till night—something the women of the country did routinely every day of their lives. Our drinking water had to be boiled, and I became adept at cooking rice with beans or lentils, which became our staple diet.

Then one day, a woman wearing a single amber bead threaded on black cotton string around her slender neck introduced herself at the gate as having been sent by Diriye to help us. Unwinding the gauzy cloth covering her wrap, which was tied to expose one bare brown shoulder, Halima gracefully went about her household duties and we grew to welcome her presence.

Diriye had boasted about Somalia's isolated beaches. One afternoon, when the children were free from classes at the International School they attended, we left E-mutt, the little dog we had found on a city street that the children had happily taken in, and piled into the Land Rover. Cole drove carefully past young children herding goats—afraid a kid might separate from the flock. Flies clustered around the nostrils of a pert little girl with a wild tangle of hair over dark eyes. Her feet danced on the hot sand like a nimble duiker, of which there were very few left in the countryside. Catching sight of our own three children seated in the back of the car, she laughed, showing chalk-white teeth, and ran alongside us. It was impossible not to fall under the irresistible charm of that little one. Exhausted, she collapsed in a heap on the sand and watched us drive on in a cloud of dust, raising an arm in response to our waves. Over dunes of soft sand, we made our way near lines of burdened camels, through a deserted village perched above the churning waves. Over hard-baked earth and more drifts we traveled, until, over the brow of a small hill, we could survey a distance so great that the shades of sea and sand merged with the hazy sky. With shouts of sheer pleasure, the children hurled themselves from the confines of the Land Rover and joined the wind that swept the sand into whirls, rolling and dashing down the slope to the inviting water below. We had been told that Shark's Bay was fittingly named, and alerted the children to the swift dark shadows in the deeper blue. Splashing in pools with brittle coral, we watched sea creatures show fragile tendrils and intricate color schemes.

Gentle waves lapped against my body as I watched Jhorna, who, with a throaty laugh and swinging black braids, leaped onto the dune just above Jessica and Sanjeeb. Sand flew and three joyful bodies tumbled over one another. Sanjeeb emerged first and dove into shallow water, trailed by his two sisters—sand sculptures with laughing eyes. When the sun shone directly overhead, we chewed hungrily on bread wrapped around chunks of tuna fish and a wedge of raw onion. Peeling back the thick skins of grapefruit, we enjoyed the sweet juice that trickled down our chins. Then, as feathery clouds changed patterns in the sky, we spread mats in the shade of a twisted tree and rested, before it was time to make tracks back to the urban side of things.

Leaps and Decisions

One clear midnight, the throb of celebration roused us from our beds and we walked down the hill to join the throngs and surging spirit in a Mogadishu square. Under the rustle of dry leaves, men, straight and lean as arrows, leaped in unison as if on balanced springs, each bounding higher than the other. Low rhythmic humming and clapping accompanied their movements, and the children stood transfixed near the center of the action. Pushed forward by the crowd, they caught the exhilaration of the occasion. Now and then, a woman joined in with the men, who raised their sticks and clicked their heels.

Much later, a full moon shone on Mogadishu as we made our way up the hill, back to our villa. And long after the children had gone to sleep, we sat under the moon's light to talk about Cole's dissatisfaction with the work he was assigned to, which had culminated with him writing a letter of resignation to UNICEF headquarters. As he talked, I noted the plants were just beginning to blossom in the variety of containers I had planted them in and balked at the thought of leaving our hilltop. But knowing full well Cole would not tolerate a job he was not satisfied with, I leaned back in my chair and said, "I never thought I would have enjoyed living in Mogadishu as much as I have."

The next morning, Cole met with the visiting Regional Director, Karl Eric Knuttson, and the Deputy Executive Director of UNICEF. Not certain as to what the outcome of their discussions would be, he ran into Liv Ulmann, UNICEF's goodwill ambassador, who was seated in the lobby and equally distressed with the Somali office's muddled arrangements for her to visit projects the organization was associated with. Welcoming Cole's invitation to come to our home, the beautiful Norwegian actress looked over the wall of our courtyard to see neglected children press dirty rags to their noses saturated in petrol, the excess of which they took turns drinking from bottles—something that left us with a terrible sadness for the situation that led to their actions. Touched by the plight of the children, and still annoyed with the Somalia office, Liv flew back to New York and shared the interview she had taped with Cole with the executive director. Soon after, Cole received a telex from James P. Grant, inquiring whether he would consider working for the UNICEF program in Uganda. Burned once, Cole was twice shy and requested a visit to Kampala before he accepted the offer. The request was denied.

It was during the month of December, and we traveled as a family to Kenya to celebrate Christmas. While Cole drove to Uganda to investigate the situation, the children and I camped in the Mombasa sun. Early every morning, local fishermen caught the waves of the Indian Ocean to surf their boats back to shore and give us a choice from their catch of the day. Other vendors brought fresh produce strapped to their bicycles in baskets. Now and then, an ocean storm washed over our tents, but the skies soon cleared. Tide pools full of sea creatures occupied the children, who easily made friends among the other campers, leaving me time to sketch and read until Cole returned to tell us that, although the office was a mess, he would like to take up the challenge in Uganda. He then reported to the UNICEF Regional Director in Nairobi that he had visited Kampala and told Knuttson that he would accept the position only if they transferred the representative and made him the officer-in-charge.

Confident we would soon have the travel documents we needed, we returned to Mogadishu to pack our personal effects and take leave of E-mutt, whose small eyes flecked with yellow looked adoringly at me as I patted his scruffy neck. Giving him one last hug, I left our little dog standing on his short brown legs, content that E-mutt would remain on the compound with Halima, who would continue to work at the villa. Months later, the occupants of our home wrote to our anxious children and claimed that our dog's bark had become deeper and that he guarded the premises well.

The last to board the flight from Mogadishu to Nairobi in December 1980, I watched the jagged coast of Somalia below, and reflected on the peaceful days we had known in the Horn of Africa. During our stay in Somalia, local children still obeyed their elders, and citizens spoke proudly of their cultural traditions, unlike the lawless situation that was Somalia in early December of 1992, when the Americans launched an operation known as Restore Hope. While helicopters roared and camera crews chased American marines up the Mogadishu beach, the world watched on their television screens as amphibious armored vehicles rose out of the Indian Ocean and rolled up to the airport with troops ready for action. The Somalis, who had avoided the area for two years because of the civil war, milled around the tarmac and were eager to shake hands with the Marines, whom they regarded as their saviors.

But mayhem continued, and up to this day, marauding thugs high on qat commandeer Mogadishu's citizens in "technicals" loaded with machine guns, anti-tank rockets and grenade launchers. Clan fighting took the lives of civilians in the central town of Baidoa, and the UN was forced to evacuate their aid workers from the southern port of Kismayu. Following further chaos and the shooting down of the helicopter that formed the central action in the film *Black Hawk Down*, the situation deteriorated and demands mounted in the United States and other countries for the withdrawal from what critics saw as an unwinnable quagmire.

Despite bitter rivalry, Somali warlords agreed to a no-shoot order on light aircraft filled with the leaves and shoots from the *catha edulis* bush. This was qat, the popular "green gold" Kenyans sent to Somalia on chartered airplanes and which earned them millions of shillings every day. The flights landed on the airstrips of the various warlords they were delivering the natural amphetamine to, who compensated the pilots well for their risky ventures. The habit had been banned in 1983 by Barre, but this did not deter the qat-chewers. Barre died in exile in January of 1995, after being forced to flee from Mogadishu to Nigeria, where he had been granted asylum.

Cole revisited Mogadishu as UNICEF's Regional Director for East Africa in mid-1992. "The American flag was hoisted within the walls of the American Embassy compound," he told me following his visit with Jonathan Howe, who he found seated in a metal container that was his office. Confident of his mission, the retired American Admiral was isolated and absolutely paranoid about security. And while they talked, helicopters clattered overhead in a quest to blow up guerrilla targets and drop commando squads in the ongoing search for the elusive Aideed—the warlord who played a major part in bringing down Mohamed Siad Barre. "The future looks rough," Cole concluded when he reached home.

Shortly after Cole returned, the late Audrey Hepburn visited Somalia as UNICEF's goodwill ambassador. Very much affected by the plight of the Somalis, particularly their children, she referred to their miserable condition in their war-ravaged land as tragic situation beyond compare. Audrey was no stranger to suffering: she had shared her story of growing up in the Nazi-occupied Netherlands when Cole and I accompanied her and her devoted companion, Robert Wolders, for two weeks in 1989 to visit UNICEF-funded projects when Cole was the representative in Bangladesh. Presenting her trim figure in bold colors that caught the photographer's eye, Audrey spread her generous smile and repeatedly captivated her audience, whether they were dignitaries, villagers or those of us who were convinced of her dedicated concern. One day, as we approached a village, she stooped to talk to a child who sat separately from the others under the shade of a coconut tree.

"Of course there's a reason why you haven't joined us," Audrey said, picking up the little girl, whose legs dangled uselessly—crippled by polio. That day she won my heart.

Audrey and Robby arrived early the night they came to our house in Nairobi for dinner following their flight from Somalia in 1992. Hearing the car drive up, I was hurriedly selecting a blouse to wear when Audrey quietly entered the bedroom and closed the door behind her. "Am I welcome?" she asked, hugging me close. "Has it really been three years since I've seen you?"

Curling up on the chaise longue near the window, she stared thoughtfully at the jacaranda trees that dropped purple flowers in the late afternoon light. "It's so lovely," she whispered, and suggested we trade houses for a holiday. Then, turning to me, Audrey grieved. "How to endure such heartache, Marilyn—those children suffering so terribly . . . dying in the camps. Starvation literally stared me in the face," she continued as she turned toward the tranquility of the garden. Time ticked on as our guests waited expectantly for the popular actress—but still we continued to discuss matters that concerned us both—the plight of children; and faith, and if there was comfort facing death.

Audrey then confessed she had not been well, and had been passing blood. Only a few months later, she died of cancer at the age of sixty -three. With sorrow, I reread the cards she had sent us and concurred with the tribute David Ansen's tribute in *Newsweek*, describing Audrey as "a princess in disguise" with a "timeless elegance" that defined her time.

But what Ansen had not mentioned was that the world had lost an irreplaceable voice for children.

Part Two

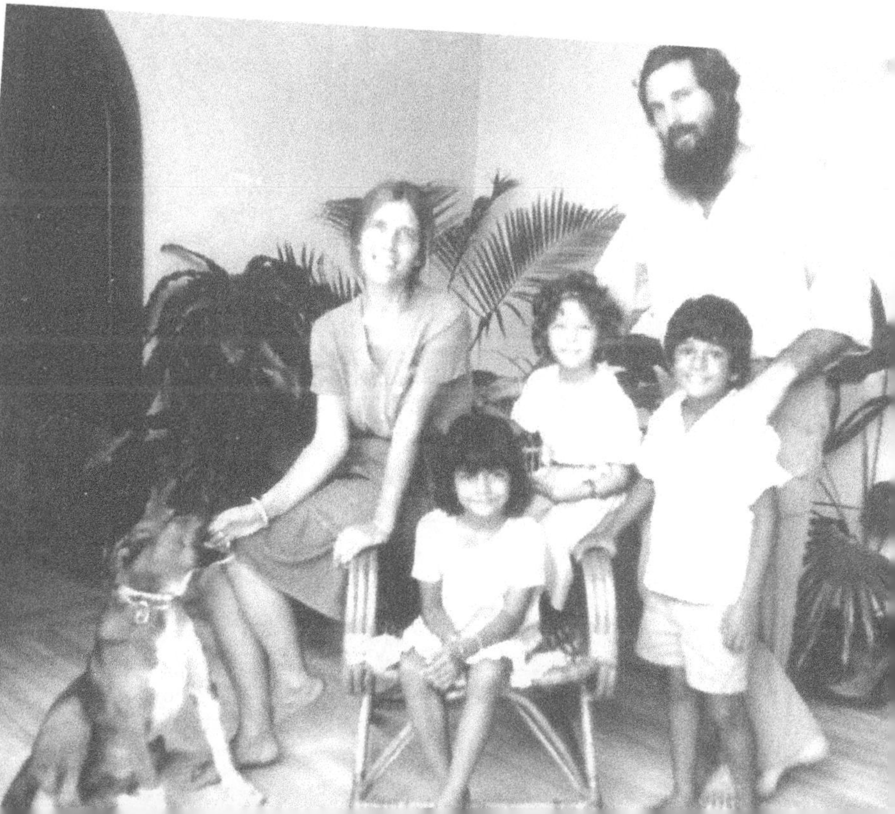

Chapter X
A Garden of Sunshine and Deadly Nightshade*

The Journey

IT WAS THE SUMMER of 1981. White and pink frangipani blossoms lay scattered on the driveway and I picked up a couple to take with me, before taking one last look at the cottage the children and I had lived in for six months. Joseph, who was waiting for me in the car, soon caught up with Cole and the children, who had left a few minutes earlier. Blinking his headlights at the Land Rover to let him know we were behind him, Joseph followed Cole toward the escarpment through heavy fog. But over the hill, the mist lifted and the magic of the African dawn merged the panorama into muted shades as far as the eye could see. Far below, Mt. Longonot's cratered peak was silhouetted against the pale sky. Deep creases shadowed the sides of the crater, where lava had flowed from an eruption many years ago. Farther along the grassy plains of the Great Rift Valley, which extended 3500 miles from the Afar Depression in Ethiopia to Mozambique, a lone Maasai herded his cattle and goats. The blanket he wore over his shoulder was a dash of red against the golden savannah. Zebra grazed under acacia trees that spread their lacy branches and giraffe bent their graceful heads to browse on low bushes. In the distance, flocks of flamingos lined soda-lake shores with crescents of deep pink.

* * *

---

*The title is taken from Winston Churchill.

Trucks with fuel bound for Rwanda and Uganda slowed us down under jacaranda trees that bordered the main thoroughfare in the town of Nakuru. Joseph followed Cole onto a side street and parked the Volvo beside the Land Rover near an Indian tea shop. Nearby, street children who had been scavenging through piles of rubbish hurried over to ask us for a shilling. Left to their own devices, the neglected children who called the pavement their home survived on food scavenged from garbage cans, petty theft, begging and prostitution. The glue they sniffed put them in a kind of stupor, which they readily admitted helped them cope with the hurtful circumstances they encountered on the street.

The hungry boys were wary of the *askari* who threatened them with his stick. Dividing the chapattis and shillings we gave them, the boys thanked us and wished us God's blessings—words of appreciation we had to remind our own children to say. "*Safiri salama, kwaheri,*" the boys shouted as they raced down the street after us, until, out of breath, they reluctantly fell behind.

The road we followed led over hills and through acres of manicured tea estates. Tea pickers wearing yellow raincoats worked their way through the aisles of tea bushes. Some carried their baskets to a small shed where they poured the leaves they had picked into piles. The small concrete houses they lived in dotted the hillside close to the town of Kericho. Our destination that night was the Tea Hotel, which stood at the top of a hill with a view of the hillside covered with tea. It was crisp and cool at the high altitude. Fog wrapped a blanket around us as we unpacked our bags and hurried into the hotel, where we asked for a fire to be lit in our room. The next morning, waiters served us pots of Kenyan tea and a hearty breakfast of eggs, sausage and bacon. "There's no hotel in Kampala that will offer us a meal like this," Cole told the children as they spread butter on their toast. What we had not eaten, I packed into a brown paper bag to eat along the way. A light drizzle obscured the view from the Kenyan highlands as we drove on until, as we descended wooded hills, sunlight captured the sheen of Lake Victoria—and far beyond, the banana green of Uganda.

Warm wind pushed my hair back from the open window as vistas changed from thorn trees shadowing the plains to fields of sugarcane and maize. Not far from the lakeside town of Kisumu, a young boy from the Luo tribe—whose people were said to have followed the River Nile south to find pasture around Lake Victoria—prodded his livestock with a branch of hibiscus bright with orange flowers. Across the road, boys of the same age group wore dark green school uniforms and chased each other on the sports field.

"These boys are fortunate to be able to go to school," Joseph mused. He was a member of the dignified Batooro tribe in Uganda, and regretted that so many of the children did not attend classes in his village. "It wasn't like that before," he continued. "Children used to wear bright uniforms when they went to school to learn about our history, traditions and values, but now . . ."

Shifting his position in the driver's seat, Joseph braked in order to ease the Volvo over a rough stretch of road. "The first potholes," he said, commenting that Kenyan roads were far better than those in Uganda. Ahead, a piece of cardboard propped up against a rock had "Help Volunteers" written on it. One of the boys beside it approached the Volvo with a wide grin and extended his hand for coins. The other boy continued breaking a pile of stones with a large rock to fill the holes in the road. Overhead, massive thunderheads reared, in contrast to startlingly white feathery clouds. Tangled barbed wire supported bougainvillea that splashed sprays of deep pink at the side of the road near trees that dripped green mangoes—colors too vibrant for canvas.

Behind us, a *matatu* driver honked repeatedly to allow him to pass. The conductor balanced precariously on the step of the sliding door waved spiritedly as they lurched past, then ducked his head into the vehicle congested with passengers. Farther down the road, blue-uniformed police pulled the *matatu* over, most likely for *kiti kidogo*, a little change.

The Kenyan border town of Busia bustled with activity. Women with children tied on their backs with *kanga*s swept refuse into piles in front of shops. Bicycle repairmen and hairdressers made brisk business under the shade of large trees. Truck after truck lined the road, with drivers accustomed to long delays sprawled under their long-haulers. Traders displayed everything from porcelain cups and enamel plates to belts, handkerchiefs, flashlights and lanterns for sale or cross-border trade.

"We have arrived," Joseph said, as he pulled the handbrake and parked near the Kenyan immigration office. "It shouldn't take you long," he said pointing to the office I had to enter to have our exit declaration forms and passports stamped. Joseph and Cole disappeared into other buildings to take care of the vehicle documentation.

A hundred meters farther on, we left the relative peace and predictability of Kenya. The ambiance was noticeably different as soon as we crossed the border into Uganda. Barefoot women with baskets of produce on their heads avoided the main gate and made their way through a broken fence. But the man struggling with his wheelbarrow loaded with maize had no choice but to pass the officer who took time to open the gate.

"That boy will have to forfeit from what he has," Joseph told me, waving aside the women who raced toward us with armfuls of pineapples. Others seated on mats beckoned us to buy their tomatoes, onions and cabbages. Moisture hung in the rain-washed air.

Cole and Joseph were familiar with what was required and walked to where they filled out vehicle registration forms. The children remained in the Land Rover to comfort the cats traveling with us, while I walked up the steps to the Ugandan immigration office. There was a strong odor of rats in the dark room. The officer seated near the light of an open window behind a large desk stifled a yawn. Folding the newspaper he had been reading, he inquired of the man who pushed ahead of me, "How long will you be in the country?" Motioning for the man to step aside for further questioning, he idly reached through the metal bars for the entry cards I had filled out.

"You plan to spend time in Uganda?" he asked as he thumbed through the pages of our American passports to stamp the visas.

Nodding, I informed him Kampala would be our home.

"You will live there?" he inquired, looking at me intently. "You know, it is not a secure place. "Safe journey," he added, as I gathered up the passports and walked out the door.

At the foot of the stairs, a small barefoot boy held up a burlap bag that dripped ice water. "*Mzungu,* you buy soda?" he inquired. Unwilling to take no for an answer, the boy and his friend, who wheeled a car he had ingeniously wired together with a Kimbo lard container, bits of rubber and tin—complete with steering that turned the wheels made from plastic lids—followed me through the parking lot. Loud voices emanated from the customs office, where Cole pushed through the crowd toward the counter and a surprisingly unperturbed officer, who handled his paperwork. There were notices tacked to the bulletin board on the wall near the office and I turned the pages of a faded calendar picturing the Source of the Nile and Murchison Falls. *Tourist destinations,* I thought, as the young man near me remarked it was no longer common to see foreigners traveling to Uganda.

Moving on, the man flicked through wads of Uganda shillings worth thousands to the dollar and approached a truck driver who had a towel thrown over his head. The Somali ignored the money-changer and shouted impatiently at the turnboys loading boxes of goods in the back of his vehicle. Stuffing more *qat* into his mouth, which was already distended with the leafy narcotic, he brazenly ogled the good-looking woman who swayed by. Other women lowered their eyes and loitered accessibly near the truck drivers who patronized their services.

The woman near me called my attention to the basket of bananas she balanced on her head. "*Bagoya?*" she asked hopefully.

The child tied to her back with a colorful *kanga* smiled and reached for the shillings I handed her mother. Peeling one of the bananas, I threw the skin to goats scavenging in a heap of rubbish and walked back to the car.

Our children were restless in the humid midday heat. It had been almost two hours since we had arrived on the Ugandan side of the border.

"What's taking so long, let's go!" Jessica and Sanjeeb grumbled, tired of reading and eager to get on with the journey. Activity at the border had slowed almost to a halt. Even the traders rested in the shade of buildings.

After a while, Cole and Joseph appeared, to drive us to where men sat lethargically astride their bicycles. The officer who emerged from a small shed ordered the men aside before she carefully scrutinized the vehicle's logbooks. Handing them back to Cole, she cupped her hands over the window of the Land Rover and looked expectantly into the back of the car.

"Anything for me?" she inquired.

Satisfied with the banana I gave her, she pushed open the gate and waved us on toward rows of unpainted concrete buildings with boarded doors.

The Cycle of Life and Death

Sudden showers washed the impossibly green countryside as we drove into Uganda, a country that in later years would be thought of by the foreign press as "a beacon of hope for Africa," but now was in a tragic state.

Patches of sunlight exposed sections of cut forest where age-old trees thick with tangles of creepers and vines soared into the blue. Smoldering ash reduced hardwood to bags of charcoal propped up for sale like sober warning sentries alongside the road. People needed money. Charcoal had a ready market and truckers made a little on the side by purchasing a few bags and selling it along the way.

On the outskirts of the town of Iganga, men planed wood on layers of sawdust to hammer together hastily built caskets to bury the dead. Deaths were frequent in Uganda. Stories abounded of killings that continued through the years, such as the time decaying burnt corpses were unearthed from pits on the side of a grassy hill in Rukungiri District, southwestern Uganda, or were found thrown into latrines and under concrete floors in buildings in what Uganda's *New Vision* daily newspaper said was the single worst incident of cult killings in world history.

Those said to be responsible for the carnage—Joseph Kibwetere, Credonia Mwerinde (who told of visions from the Virgin Mary) and Dominic Kataribaabo, an excommunicated Catholic father—had deliberated on how to halt the imminent end of a sinful world. Linking their vision to begin a movement that would enforce strict adherence to the Ten Commandments of God, they said, "The Catholic Church is backsliding. It offers no solutions to our problems."

Rev. Fr. Paul Ikazire joined the secretive Kanungu movement in his own protest against the organized church. "At first we had good intentions," he admitted. "Priests were covered in scandals and the Aids scourge was taking its toll on the faithful. The world seemed poised to end," the priest told James Mujuni in an interview reported in the *New Vision* (not unlike what fundamentalists had predicted in other parts of the world at the end of the millennium).

In time, Ikazire became disillusioned with the sect, and on Sunday, 27 February 1994, the eighty-three-year-old priest led mass at his house and boldly advised the Kanungu members gathered there that the way they had chosen was very narrow.

"*Orurite embwa rugigaza enyindo kwonka orurite omuntu rumwigazaakanwa,*" he was quoted as saying in the *New Vision*: "The death that will kill a dog first impairs its capacity to sniff but the death that will kill a man first seals his mouth."

"The rule of silence is a violation of our rights," the frail old man told the parishioners. "Having to sell our property is designed to impoverish us . . . make us more submissive." Convinced then that it was time to quit what was known as the Kanungu cult, Ikazire warned the congregation to follow him and "never return." Had the priest not left the movement, no doubt he would have been yet another victim of the grim mass murder.

Uganda's daily *Monitor* continued to examine the horror of Kanungu, and on 28 April reported that fewer Ugandans turned up for Easter mass in 2002 than was usual. Increasingly wary about the church and its practices, those asked to bring candles to worship responded: "That's how Kibwetere burnt people." More than three hundred people were reported to have died in what appeared to be mass suicide by setting themselves on fire in the Church of the Restoration of the Ten Commandments of God.

Although religious psychological manipulation was not new to Uganda, Kanungu was yet another truly disturbing event in a country that Winston Churchill declared in 1907 in *My African Journey*, "must be too good to be true . . . Behind its glittering mask (Uganda) wears a sinister aspect . . ."

* * *

Ahead, a twisted scrap of metal obstructed the road to Kampala, one of ten roadblocks we would encounter that day. Joseph slowed down and stopped the car not far from army men lounging against a termite mound. One of the men stretched lethargically and sauntered toward us.

I had heard many tales about Uganda's disreputable army and lowered myself in my seat, just as I had done when a drunken soldier in Nigeria had lurched out of the bush and demanded we get out of the car. Uneasy now, I turned to make sure Cole was following in the Land Rover.

The soldier took his time to light a cigarette. Inhaling deeply, he adjusted the strap that held his AK-47 and glanced at the Volvo's license plate. "CD96," he muttered, before he spotted the blue UN flag hanging limply on the wing of the car.

"Move on," he ordered curtly.

"Not so for us civilians," Joseph remarked soberly, as he eased the car around the rubber tires put there to impede traffic. "If you cannot offer shillings . . . you know how it is," he said, assuming I did. "These men control with the gun. They can threaten you until their needs are met."

Aware he had his own experiences to tell, I asked him about his past.

"My parents planted seven acres of land in the Kingdom of Toro with bananas that we used to make *pombe*—local beer my mother sold at a small bar," he began, as we followed the road bordered by crumbling buildings, where barefoot men assembled in front of shops bare of merchandise. "In later years I moved to Kampala, and in 1969—"

"Wasn't that the year when there was an assassination attempt on President Obote?" I interrupted, offering what I knew of Uganda's troubled past.

"Yes," Joseph said. "It was then that I met an embassy officer at the golf course where I used to carry his clubs. He offered me a job. That man trusted me as a *shamba* worker in his garden. He left me caring for his house, and when he returned to Uganda, he brought me things that almost cost me my life. Soon after, thieves swung a bag of bricks at the door and forced it open. Then they cut me with a *panga*. They tried to kill me for what they wanted," Joseph remembered, fingering the scar on his cheek. "I shouted for help and when my neighbor came, they smashed his head and he died. My boss left for America. Before he went, he advised me to take up driving."

"And then you found a job with UNICEF," I said, completing his story.

The truck ahead spewed black diesel smoke on the elephant grass that grew beside abandoned roadside plots. Flooring the accelerator, Joseph passed the dilapidated vehicle loaded with maize, the staple crop of eastern Uganda. Farther on, half a corroded metal barrel blocked the road before another soldier flagged us down. This time, the man in camouflage uniform bent his head close enough for me to smell *waragi* on his breath.

"You are going where?" he slurred.

"To Kampala," Joseph replied tersely—audaciously, I thought, and ordered him to step aside. Absurdly, the line from an old song: "*A lot of men didn't and a lot of men died,*" circled in my mind. Defiantly kicking the tire of the Volvo with his unlaced boot, the soldier said, "You go—go!" and meandered over to where Cole had parked behind us.

The children had their faces—brown, white, brown—pressed against the windows of the Land Rover. It troubled me that their introduction to Uganda was with drunken soldiers.

Joseph stopped the car in deep shade in the Mabira Forest, near a sign with PUBLIC PICNIC SITE barely discernible on the worn wood. The steady chorus of frogs filled the damp air as I walked over to the Land Rover Cole parked behind us. The children were ready to stretch, and clambered over the seats and our two unhappy cats. Meowing disconsolately, the cats arched their backs when I reached through their cages to stroke them.

Tall reeds grew in the damp earth beside the road, surrounding the towering brown castles of termite mounds. I watched a pair of hornbills land noisily in the canopy above us. Leaving the children with Joseph to watch the distinctive birds, I slid onto the front seat of the Land Rover beside Cole.

He was tired and bent his head in a gesture I knew meant that he wanted me to rub his neck. Idly pushing his finger into one of the jagged holes scarring the dashboard, he said, "Bullets caused those."

"So what happened?"

"About two weeks ago," he said, "Paddy O'Hanlon, the UNICEF supply officer, closed the door of his Renault 16 and turned on the ignition to drive back to his residence." Minutes later, Joseph, who was driving twelve staff members on a shuttle from the office toward Makerere University, was startled to see armed men driving O'Hanlon's Renault. He gave chase, unaware that the hijackers had accomplices. The bandits circled Wandageya roundabout alongside Joseph and shot at the Land Rover's windows, spare wheel, bonnet and radiator.

"This car?" I asked, shifting uncomfortably on the seat the terrified passengers had been sitting on. "Was anyone killed?"

"Miraculously, no one, but Ochama was hurt," Cole said. "He had glass lodged in his eye. Thankfully, Joseph managed to control the vehicle and to this day he prays with thanksgiving on the anniversary of his survival."

"What about Paddy?"

"He walked back to the office pretty shaken, but unharmed."

"Did the men driving the Renault escape?" I asked.

"For the time being. Then *my* drama began," Cole replied, knowing I would persevere until I had heard the end of the story. "Early the next morning," he continued. "Obote's vice president, Paulo Muwanga, called me to his office and told me it had not been necessary to evacuate the UNICEF staff to Kenya." But Cole had stood his ground. "Security has collapsed," he told the vice president, and added that no less than thirty-five diplomatic-registered vehicles had been stolen in the city in broad daylight.

Concluding that the office would remain closed until the Renault and three UNICEF vehicles were returned, that evening Cole repeated the same message in an interview with Jimmy Dean that was broadcast over Radio Uganda.

Again Muwanga summoned him and told him he would not tolerate United Nations officials issuing anti-government statements on state-run radio and television or at press conferences. Aware of Muwanga's brutal reputation, Cole left the office a very worried man.

One tension-filled week later, Basilio Okello, Obote's Army Commander in charge of Kampala, rolled up the Mbuya driveway in a chauffeured black Mercedes, followed by an officer in a Land Rover towing O'Hanlon's bloodstained Renault 16.

Stripped and beyond repair, the other vehicles were returned the following day.

"After that," Cole said, "Radio Uganda reported that the government had improved the law and order situation—a situation that still is far from good. But," he added quickly, throwing his arm around my shoulder, "at least *we'll* be together."

Disturbed by what he had told me, I slid closer to his reassuring presence as thunder resonated overhead and lightning flashed through the trees. The cycle of life and death is so rapid here, I thought, as my eyes swept over the hauntingly beautiful green countryside, rich with fertile soil. Plant a seed and six plants will spring up, it was said of the productive land, which Britain had ruled as a protectorate. Bright sunshine alternated with rain showers, and the country's temperate climate produced excellent crops and plants that were in perpetual bloom. But the air felt oppressive—thick.

Dense vegetation to the side of the road trapped the warm humidity and I stared at the dark, brooding rainforest where Joseph told me that only weeks earlier undisclosed numbers of bodies had been found.

I turned to Cole. "You feel confident, then?"

He squeezed my shoulder and said, "Time to go," as bursts of laughter came from the direction of the children. This time, Sanjeeb buckled himself into the front seat next to Joseph.

"Did you put this here?" he asked Joseph, running his fingers over the small silver cross glued to the front of the dashboard.

"I put it there because, by the grace of God, I survived a shootout," Joseph told him solemnly. "There were many people in the car that day but no one was killed."

"What happened?" Sanjeeb asked. But Joseph changed the subject.

Broken tarmac slowed our journey into the town of Jinja. Joseph pointed to the side of a grassy hill. "That is where the British train President Obote's army. Can you believe this place was once a busy urban center?" he continued as we drove down the former industrial town's main thoroughfare, past broken light poles and unkempt buildings. The merchants who had taken over abandoned Asian shops had the metal doors of their *dukas* partially open. "The few wares they have they sell for exorbitant *magendo* prices," Joseph said, as we passed men seated under the shade of a verandah, who responded lethargically to our waves. As soon as we left the depressing town, ornamental flamboyant trees with clusters of scarlet flowers lined the bumpy dirt road that led toward the lake and river of legends.

"It was just there," Joseph said, pointing at the water from Lake Victoria, the second-largest lake in the world, which plunged across Owen Falls dam to the beginning of the White Nile. "There I saw people being thrown when Amin's troops were fleeing. And here," he said, stopping the car. "This is where I stopped when I was running from Kampala. That day, the Tanzanian troops were running and Amin's soldiers started to slaughter people. They threw those who were trying to escape into the river," he repeated, shaking his head in disbelief.

"And you? How did you escape?" I asked, shivering in spite of the warmth of the sun. For what he described had occurred just over two years before we arrived to make this troubled country our home. I needed no further misgivings about taking our three children to Kampala, despite having been witness to adversity and certainly being familiar with change.

Far below, the water churned into pools where the bodies had been thrown. Above the dam, fishermen threw their nets from small wooden boats next to papyrus swamps in the still water.

"Old father Nile without any doubt rises in the Victoria N'yanza," John Hanning Speke declared in 1862. Claiming the lake was the source of the "holy river," the explorer further described the expedition's discovery of the Source of the Nile in his journal as beautiful but not what he expected ". . . for the broad falls, about 12 feet deep, and 400 to 500 feet broad, were broken by rocks. Still it was a sight that attracted one to it for hours—the roar of the waters, the thousands of passenger fish, leaping at the falls with all their might, the Wasoga and Waganda fishermen coming out in boats and taking post on all the rocks with rod and hook, hippopotami and crocodiles lying sleepily on the water . . ."

There was no evidence of hippopotamuses or crocodiles now, I reflected as I gazed out the window at the water spilling over the dam to follow its winding river course 6160 kilometers through the swamps and sand dunes of Sudan and Egypt to the Mediterranean. Recovering my attention, Joseph started the car and repeated, "I will never forget these falls."

"Where did you run to?" I asked.

"I had then to take the Kiyunga route," he said. "There I dismantled the UNICEF Land Rover I had been driving. I left it like a wrecked thing at a hospital compound before my family and me fled to the bush. Eh, eh, eh, that was bad," he said softly, as he steered the car past a chassis left to rust on the side of the road. "Near my home, on my return to Kampala, an armed soldier opened fire and killed two of Amin's tribe who were drinking millet beer. It was then I chose to leave the place and move."

Disturbed by his tale, I stole a look at the expression on his face, but it told me nothing. Had people become so accustomed to violence they paid it no heed?

The road to "the garden of sunshine" was in disrepair. "Imagine, they call this a road," Joseph complained, as clay pots placed on a stump beside the road caught my attention. "Namave pots," I read off the sign.

"The former owner of that small workshop was killed some time back," Joseph told me, dampening my hopes for what might be for sale. Ahead, a dilapidated signboard had an arrow pointing to what had been the Industrial Area.

"Welcome to the Kingdom of the Buganda," Joseph said, as we drove past people hurriedly walking to reach their homes before sundown. "Here one person is known as *Muganda*, many persons, *Baganda*, and the language spoken is *Luganda*."

Kingdom? How could things have gone so terribly wrong for this once

-flourishing country rich with culture? I wondered as we passed a giant rubbish heap where a boy in tattered shorts rummaged through decomposed matter to find what he could to eat.

It was approaching dusk in the city of seven hills. Shadows were deepening. Kampala's streets were still wet from the late-afternoon storm that had swept over the city from Lake Victoria. Streams of muddy rainwater washed down the sides of the broken sidewalk into the numerous potholes. Afraid of getting wet, a woman pulled her barefoot children to her side as Joseph slowed down to avoid splashing them. Opposite the rundown petrol station, giant marabou storks probed with their long beaks in the waste that clogged the drain near a broken culvert.

"We're almost home," Joseph said as he turned left on Port Bell Road and rounded the corner past the unfinished crumbling East African post office training building. The road dipped before we could see smoke rising from fires near dozens of uniports that housed the military on the side of the hill. Broken electric and telephone wires hung from posts leaning over the steep dead-end road we turned onto. "To the left," Joseph said, his voice eager with the prospect of reaching home, for he lived on the same compound with his wife and daughters, "are the Verona Fathers and the Vatican nuncio where the ambassador to Uganda lives and works."

Fifty yards farther on, the guard pushed open the gate and I could see the house that would be our home perched on the hill.

Chapter XI
On a Rise with a View

THE DALMATIAN COLE had brought from Nairobi on an earlier trip raced beside the car as we drove up the gravel drive. Sanjeeb leaned from the window and called to the excited dog, which Graham Hancock, the author of *Lords of Poverty*, had given us. Even the cats sensed the end of confinement in their cages and arched their backs to hiss at Bender, who was dashing about the car in ecstatic circles. Pelesia, who had heard all the commotion, stood at the front door of the house clad in a bright *kanga*, her hair twisted into short antenna-like spikes. The girls were delighted to see the warm-hearted woman who had worked for us in Kenya, and raced up the stairs to give her a hug. It had been her choice to accompany us to Uganda, although she was concerned about the stories she had heard and read about in local papers.

"I am so fine now that you are here," Pelesia said cheerfully. She had come by bus a few days earlier. "But I hear a lot of gunshots," she said as she helped carry our luggage into the house she had cleaned for our arrival. "And I do fear leaving the compound," she told the children, who chorused: "But we are here and in Kampala you are closer to your village." Jessica reminded Pelesia that she would not have to travel far to reach the small town in Kenya she came from, close to the Ugandan border.

Pelesia laughed and took the girls down the hill to see the double metal uniport she had moved into. Soon, music resonated from the radio she turned up in her little house. The catchy beat attracted Sanjeeb, who joined them. After a while, they walked back up the hill to explore the surroundings that would be our home. Spotting the tire Cole had hung from a tree at the back of the garden, the children raced to swing on it. They were disappointed that the little kidney-shaped swimming pool was bone-dry, but Cole assured them he would have a truckload of water brought from Lake Victoria to fill it.

It was dusk by the time Pelesia helped me make the beds in the large room the children would be sharing. The activity of the day had caught up with us, and the children sat quietly around the table on the verandah and waited for me to unpack the hardboiled eggs and bread I had brought from the Kericho Tea Hotel. Lost in our thoughts, we shared our first meal before a view of the hills silhouetted against the darkening sky.

The mood changed, however, when, in what would prove to be an ongoing dispute, the children argued among themselves about who would be the first to take a bath. There were two bathrooms in the house, and the hot-water heaters that had been installed in 1950 took time to heat. The need to conserve water was obvious, as it had to be trucked in once a week from Lake Victoria and pumped up to fill the tank. From then on, the children opted to use the same bathwater, for if one let out the hot water, the next one had a cold bath. That night, after tucking the children into their beds, Cole and I wandered around the rooms in "the house I never made a home—just slept in," he said, as I entered the bathroom and ran water for my own bath. "Leave the water in," he reminded me as I closed the curtains.

That night I moved closer to Cole, who wrapped me in his arms when I told him I had heard gunshots. "Don't worry, they're off in the distance," he murmured, half asleep. But when a barrage of bullets sounded close at hand, I sat up and pushed aside the bedding to look out the window and check the children, who were deeply asleep with the cats curled near them. *Was I crazy to have agreed to bring the children to a country at war?* I wondered, as I crept back to bed.

The next morning, a gentle wind stirred the air, which was alive with birdsong. Not wanting to disturb Cole, I left him sleeping under the mosquito netting and walked barefoot over the parquet floors to the verandah of our new home. *No bars*, I thought, as I pushed open the folding doors to reveal a beautiful dawn still sparkling with morning stars, which cast aside my fears of the previous night. Pulling one of the weathered old chairs to face the view, I sat down to savor the moment. The air was moist and clear, and I caught my breath when a pied kingfisher hovered over the grass. In the distance, the green hills stretched toward the lake.

It had been our choice to lead a life that routinely required us to adjust to new relationships, agendas, work and homes in completely different settings. A life that offered rich cultural advantages (and sometimes disadvantages)—depending on the contrast to what we had already experienced. Sentimental about departures and the familiarity of what I had come to know, I still anticipated the new. Certainly, it benefited the family to settle into new surroundings as rapidly as possible.

Our possessions had been trucked to Kampala earlier, and this time the children were old enough to help me arrange familiar belongings here and there to create our home in Uganda. Wherever we lived, native furnishings, basketry and materials had been our decor, and now the pillows and spreads from India brightened up the low divan pushed against the wall. The walls already had nails, on which we hung pictures and posters. Much-loved stuffed animals found a place on the children's beds spread with the quilts their grandmother had given them. I claimed the old wooden desk with cubbyholes, which we shoved between windows with a view. Jhorna promptly sat down to play the upright piano we were pleased to find in the front room. Although out of tune, the piano became the hub of many a Sunday night sing we had together with whoever was living with us at the moment. Plates and cups we placed on the shelf above the enamel-topped table I found in the kitchen, which would be useful for kneading bread and rolling pasta.

Jessica soon tired of unpacking and found a corner to read Joy Adamson's *Born Free* series and other books she found on a shelf the former landowner had left behind. I was sure the proprietor would have taken the linens and crystal we found stored in a cupboard if she and her family had not been forced to hurriedly abandon their home and property not long after 1971, when Idi Amin Dada seized control of the government while President Obote was attending the Commonwealth Conference in Singapore. That day, thousands of jubilant civilians and ululating women surged onto the streets of Kampala to celebrate the overthrow.

But their joy did not last. Soon after taking power, Amin, who was of mixed Kakwe and Lugbara parentage (two dominant tribes in the West Nile), began his eight-year bloody reign of terror. He promoted himself to His Excellency President for Life, Field Marshal Al Hajj Dr. Idi Amin Dada, VC, DSO, MC, lord of all the beasts of the earth and fishes of the sea and conqueror of the British Empire in Africa. In November 1972, in a blatantly racist attack, he expelled over fifty thousand Asians who had made Uganda their home. Regardless of whether they held Ugandan or British passports, Amin gave those he said had been brought to build the railway to carry away Uganda's wealth a departure deadline and informed them they would be sent to detention camps if they remained behind. Stunned by the directive, the Asians, who were told to leave the country with a hundred-dollar allowance and a few personal effects, left behind possessions that were looted by the same soldiers ordered to protect their goods.

Around this time, the Hunts not only left their home and property behind, but also business investments that included a travel agency and something to do with the automotive industry. It was curious to me that their house and property had not been looted during the eight years they had been gone, as I knew of so many others that had been plundered.

Evenings in Kampala were often cool enough to put on a sweater. Knowing we would enjoy a fire out of doors, the children helped carry stones from the driveway to build a fireplace on the side of the verandah, which I then bordered with plants I found at a nursery near the city sewage. In time, jasmine and queen of the night scented the air. Rosemary, mint, basil and other herbs flourished in the clay pots I placed near the outdoor gazebo that served as our dining room. Mauve, coral and scarlet bougainvillea, gardenia bushes, verbena and honeysuckle grew quickly along the garden path, confirming what I had heard of Uganda's fertile soil. Lake Victoria, the second largest lake in the world, generated its own micro-climate in the form of thunderstorms that watered, often in the afternoon, what I had planted, followed by sunshine that caused everything to grow quickly and flourish.

Noisy turacos called harshly from the hibiscus bushes that were in perpetual bloom at the back of the garden. Woodpeckers nested in holes in a dead spiny branch of the bombax tree that blossomed with pink flowers. The fruit of the tree dried and split open to release delicate white fiber the children chased when it floated to the ground. Hadada ibises preened on the grass, and at dusk, a cloud of white egrets flew overhead to nest in trees. For a while, I tried a vegetable garden, which yielded cucumbers and tomatoes, but in time I surrendered the plants to the birds and insects and was left only with a few bushes of very hot chili peppers and a volunteer avocado tree.

The ten-acre vacant lot behind the house had grown into almost impenetrable bush, and one day I held my breath when a long, slender mamba slithered over the embankment across the yard. Aware that the black mamba's bite was often fatal, we kept the grass in our yard cut low, realizing that, for the most part, the snake hunted small prey such as lizards and birds, avoiding contact with humans. Another day, a mother cobra hatched her brood in an old used trailer parked near the workshop at the bottom of the hill. Pelesia, who was terrified of snakes, moaned and held her hands to her face when she heard about the incident. It took me some time to convince her that anti-venom was available (though sporadically) at Mulago Hospital in the event one of us was bitten.

Snakes, though, were of less concern than the other factors we dealt with living in Kampala. The once-thriving health services of the country were depleted of doctors, bankrupted by spiraling costs, crippled by corruption and pathetically depleted of revenues. Insecure medical stores resulted in theft, which went to the private sector because pharmacies could not import medicines as there were no foreign exchange reserves, taxes were not collected and the administration of the country was broken down. Public services, if they existed, were sporadic. What did exist was manned by unmotivated civil servants, whose salaries were often at poverty levels of one dollar per day, forcing them to plant gardens on road reserves or moonlight to earn money to feed their families. The city was crumbling, its sewage pipes clogged and broken. Half-built structures were overgrown with grass and bush. Roadblocks were common. The roads were full of potholes lined with piles of garbage, and few buses plied the streets. Even the previously spirited cultural activities at the National Theatre had come to a standstill.

Adjacent to and separated from our compound by a wire-mesh fence unevenly covered with thorny bougainvillea, military transport rumbled up the road to the congested Mbuya Barracks. Barefoot children and chickens darted about the tin sheds, which housed the unkempt, undisciplined Ugandan National Liberation Army (UNLA). Drawn largely from northern districts, the UNLA soldiers were volunteers, often with only marginal education, who worked for low pay and sporadic rations; thus, they were forced to supplement their incomes in order to survive.

The soldiers often manned roadblocks during the day and looted at night. Anyone who had money and wanted to settle a dispute or an old score, take revenge or mount a major robbery could hire the soldiers. When an army vehicle broke down, the soldiers simply waited on the road for the next unfortunate passerby, who was left shaken at the roadside with his vehicle commandeered in the best case, or shot outright if he gave any resistance. Within the rank and file of the UNLA, the command and control hierarchy had broken down, which allowed mini-warlords to prey upon those from within the UNLA who had food, fuel, luxury goods or vehicles from their own sorties into the business district, private residential areas or onto the roads.

The Verona Fathers were a buffer between our compound and the barracks. The largest Catholic mission in Uganda (the majority of whose population is Christian), the compound housed the Vatican Embassy, a church, a welfare center, warehouses, administrative offices for the entire country, a school for local children and the Verona Fathers' guesthouse for rest and recuperation. The army officer who occupied the house across the road from us kept a scruffy clutch of armed men at his gate twenty-four hours a day, and although the guards drank the local brew and could be seen staggering back to their duties, they never gave us any trouble, having been ordered to keep the peace with respect to both UNICEF and the Verona Fathers.

Farther up the hill from us, the road led past leafy residences behind high concrete walls. When time allowed, the children and I walked with Cole up the shaded path to the top of the hill to enjoy the splendid panorama of the city built on seven green hills.

In the distance, we could see the leafy green suburbs of Kololo, where the embassies and those with higher incomes were located. Old Kampala had rows of houses similar to what we had seen in India, some of which were abandoned. Buildings once owned by Asian shopkeepers had been taken over by locals; Mengo was where the historic 1900 agreement between the *Kabaka,* King of Baganda and the British colonial officers was signed, creating the Uganda Protectorate. A winding road sided with small shops led up to the Anglican Cathedral on the summit of Namirembe, and we could see the Catholic Cathedral at the top of Rubaga Hill.

Makerere, one of Africa's oldest universities, was established in 1922, but now struggled with poorly paid professors, some of whose wives baked bread and cakes to subsidize their meager incomes. Twice appointed as the vice chancellor of Makerere, our friend Professor Senteza Kajubi's explanation for Uganda's decline and broken-down state was that, after military coups had been staged and guerrilla and liberation wars waged with the purpose of putting an end to tyranny and political and economic corruption, "the succeeding regimes frequently not only followed in the footsteps of their predecessors, but went even further in abusing power, violating human rights and/or failing to promote peace, stability and the economic and welfare of the people, until they themselves journeyed the same way they came in— at gun point."

The downtown Kampala city center known as Nakasero was terribly rundown, and only a shadow of its prosperous past. Many of the shops were boarded up or barricaded with heavy iron doors to prevent access to robbers, looters or even marauding military gangs. When Idi Amin expelled the Asians, he appointed committees to allocate the companies, property and factories they left behind to hopefuls who lined up to be publicly interviewed. Displeased with the committee's insistence on qualifications, he had his army officers take over the distribution, with the result that his army staff, friends and relatives (who had little interest in sustaining or any idea how to run what they had been given) sold off their goods and plundered their stocks. Shortly thereafter, Uganda's remaining professionals fled into exile and the country's economy began to collapse.

* * *

The school the children attended was in a large bungalow (most likely abandoned by an Asian family ordered to leave the country in 1972) located on Buganda Road on Nakasero Hill, with enough rooms to accommodate the primary classes and a small flat the headmistress occupied on the second floor.

The first day of lessons, the children ran eagerly down the stairs to where Joseph was parked. Lincoln would be the fifth school they had attended. We had never considered home schooling, as they enjoyed and needed the social interaction with their teachers and other kids (besides, I didn't have the patience; teaching them piano lessons was quite enough!).

Joseph looked trim in his blue UNICEF uniform. Straightening it as he stood by the car, he welcomed the children, exuding the kind of confidence that made me feel secure about him driving them to school. Thereafter, Joseph varied his route to Lincoln School to avoid the possibility of being hijacked, which was a real risk and one of my worst fears. Incidents were common and could occur at one of the numerous roadblocks mounted by the ragtag army or by imposters dressed in army fatigues, who also mounted impromptu roadblocks for the sole purpose of robbery and hijacking. Though kidnapping was uncommon, hijackers often sped away with passengers still in the vehicle, dumping them off at some remote corner of the city or on a road leading out of Kampala.

That afternoon, when the children returned from school, they were eager to report that there were forty-three children enrolled. Then they handed me a note from the principal requesting parents to send blankets, biscuits and drinking water to school in the event there were disturbances that would prevent them from returning home. *What a way to begin the school year*, I thought, as Sanjeeb announced that the teachers had gathered the children together to tell them what to do in the case of an emergency. "At least if we can't come home, we'll have something to eat," he said earnestly.

Pleased to be home, the children put down their backpacks and ran outside to play with the animals. It was a real advantage to have a compound large (and safe) enough for them to find things to do, as organized school activities were infrequent and we abided by a curfew that kept us home in the evenings.

Ours were the only UN children to attend Lincoln in 1981. Most of the other students were from expatriate families wealthy enough to manage the dangers, deprivations and difficulties they encountered. Dr. Rick Goodgame (whose wife had grown up in Eastern Uganda and who had returned immediately after Idi Amin had been overthrown), dropped their four children at school before he drove to Mulago Hospital, where he taught medicine at the Medical School. There was also a fragile and small business community who sent their children to Lincoln, such as the Greek family who had a long history in the country with only remote ties to their native Greece; and, finally, a few kids from small embassy staff. Among the teaching staff there was a Congolese refugee; a local Asian from a prominent family; Mrs. Otto, a missionary; two others who stayed on, having worked at the university—and the principal.

In time, the UN lifted the ban on dependents, and within a year of our move to Uganda, Cole hired Thomas Ekval to head the water and sanitation. When he and his wife moved to Kampala, they enrolled their two sons in Lincoln. Cole had met Thomas in the Somali region of the Ogaden when he was the Oxfam Field Director for Ethiopia. "He was camping in a tree to avoid the bugs and hyenas," Cole told me after he had spent time with Thomas, observing the work he was doing with hand-dug wells.

In this way, the school gradually grew to one hundred students and more by the time of the Okello coup in mid-1985, after which enrollment again shrank to about sixty children.

One day, anticipating the prospect of unexpected time at home to do what they liked, the children slammed the doors of the Volvo and ran up the stairs to the house. "There was a shoot-out near Wandageya and Papa sent Joseph to bring us home!" Sanjeeb shouted. The Wandageya roundabout he referred to was the scene of many such incidents, as it was on the main artery to northern Uganda via the infamous Bombo barracks and the Luwero triangle.

"I'm glad you're home," I told the children, having been affected by the same gun battle while driving a friend with physical disabilities to town. The afternoon had been quiet enough until suddenly, over the brow of the hill, crowds of people surged toward us like waves over sand and I panicked. To make matters worse, my friend gripped the dashboard and kept telling me to "look at the people who must be running from some shootout."

"You must reverse," he said, just as bullets smashed into the concrete wall of the two-story building to the right of us.

Terrified, I watched people scatter in every direction. With nerves taut as a bowstring, I sat as low as I could behind the steering wheel and continued driving until we reached the dip that turned off the main road. Greatly relieved to have left the chaos behind, my friend sank back in his seat and said, "All of this is surely not a good thing."

"It's terrible, this fighting," I said weakly, parking on the side of the road in order to recover my wits.

Before we moved to Kampala, my friends had confronted me, wondering whether I could cope with gun-toting military. "Of course it won't be easy," I told them. But, having dealt with insecurity and incidents, I was convinced I could deal with the "edge" of living in Uganda. Although one almost anticipated the daily stimulus and some people became addicted to the adrenalin rush, I loathed running into people and situations gone amok and found my emotions fluctuating between enjoying our dignified Ugandan friends and the country's natural beauty and being upset by the wretched conditions the country and society had fallen into. At the same time, I recognized and was reassured by the privilege of having a guarded home and compound, Cole's work, connections, Lincoln School, the commodity runs to Kenya and the ability to move in and out of situations—and the country, if conditions worsened. In reality, we were far removed from the poverty and severe trauma people endured, but in a sense we could learn from their endurance and, even if it was from a position of privilege, sympathize with their circumstances and give a little of ourselves. But dealing with the disparity was definitely not a simple task . . . nor had it been in any of the countries we had lived in.

As suddenly as the disturbances had begun, Kampala's streets became quiet. Light rain settled the gun smoke at Wandageya as I drove my friend home. Thanking me profusely when I insisted on dropping him close to where he lived, he called after me to "please arrive home safely." The shootout had left me in a disturbed state, for not only were my nerves frazzled; I was also concerned about the security of the children. But when I arrived home I found I need not have worried, for the headmistress had kept them at school when the shooting started.

That evening, Cole returned from the office ahead of the curfew that allowed our family to spend evenings together. In the distance, we could hear sporadic gunshots punctuating the night. "Popcorn," as we called gunshots, was a common evening occurrence, and it was our custom to gather around the table under the outdoor gazebo to give thanks for the day's protection. It was not yet time for the eight o'clock radio check that routinely confirmed all UN international staff members were safe and sound. Watching the children scrape the last bit of pasta from the serving dish, I made a mental note to order more provisions from UNICEF's supply officer, who brought essential commodities for the staff from Kenya on the third week of every month. The commodities included powdered milk, oil, bags of flour, cereal, tinned goods and drinks.

The supply officer, Paddy O'Hanlon, had been involved in the Karamoja famine in the early 1980s, which was so severe and tragic that it attracted over twenty international NGOs—before he applied for a job with UNICEF. The relief efforts came to a close in 1982 with the arrival of normal rains. Since it was difficult to find people to work in Uganda, Cole welcomed Paddy's application and hired him as the supply logistics officer. Paddy set up a commodity run. He accompanied the truck in another vehicle (both were registered with diplomatic corps license plates) to pick up provisions for the staff in Eldoret, western Kenya, about six hours' drive from Kampala. For a couple of years the run went smoothly, until one day the Ministry of Foreign Affairs asked Cole why UNICEF was importing expensive electronic goods. Cole investigated the matter, only to find O'Hanlon was in cahoots with businessmen in the town of Eldoret in a neat little scam. O'Hanlon simply organized shipments so they arrived in Entebbe a few days before his scheduled commodity run. He then received the goods consigned to UNICEF at Entebbe Airport, which had been ordered and paid for by Asian businessmen in Kenya. The load of television sets needed no documentation, since the vehicle was diplomatic and theoretically empty. Thus the merchandise entered duty-free across the border to Kenya. Embarrassed when Cole presented him with the evidence he had gathered, Paddy resigned. He packed his suitcases overnight and left the following day.

Although very few commodities were available in Kampala, the local market had excellent produce, and once or twice a week I drove into town to pick up tomatoes, onions, eggplants, potatoes, cabbages, greens, sometimes mushrooms, and whatever fruit was in season, including mangoes, passion fruit, bananas and pineapples. It concerned me that chickens and ducks squawked from filthy feathered cages close to the rows of stalls where the sellers, who were mostly women, sold their goods. As soon as I parked the car, young boys in tattered clothing ran from where they had been sorting through piles of putrid rubbish to carry my baskets for a few shillings. School was not an option for the boys, and whatever they could earn or glean at the market helped them at least with something to eat.

The produce I purchased in Kampala in the early 1980s was grown by smallholders on subsistence farms. Salaries were so low and jobs so scarce that urban families typically staked out roadside verges for planting everything from staples—especially cassava and yams—to a variety of vegetables. This not only provided them with the food they needed, but surpluses could be sold along the roadside or at Nakasero market. David Alnwick, a UNICEF nutritionist, carried out an interesting study on urban children in Kampala with the surprising conclusion that, even during the civil war, Ugandan slum children were better nourished than their counterparts in neighboring Kenya. Increasingly aware that economics alone was not enough to either describe or understand the real situation in the country, Cole brought together a group of academics to discuss the issue. One of them was an agricultural economist, who later wrote: "The Ugandan economy has undergone massive structural shifts, in the last decade. . . . Despite . . . huge upheavals, the country's food situation remained healthy."*

**Beyond Crisis: Development Issues in Uganda*. Paul Wiebe and Cole Dodge, eds. MISR, 1987.

It was late morning by the time I had selected the produce we needed for the week. This time, I had brought my eldest sister Esther, who was visiting us from India. She was talking with a group of boys, when I heard her exclaim, "Marilyn, someone snatched my gold chain!" The traders immediately stood up from behind their stalls and shouted loudly for someone to stop the thief. Fortunately, it was a youngster who caught up with culprit, who dropped the necklace and ran—or we might have witnessed mob violence.

After purchasing produce at Nakasero Market, I had planned to take Esther to see the Kasubi Tombs—the largest grass-thatched hut in the world, constructed in 1882, where four Buganda kings were buried, but we had spent more time at the market than we had intended and decided to go straight home. Pelesia heard the car drive in and stopped the washing she was doing by hand in the old bathtub on the back verandah to help us carry the baskets of produce to the house. Esther then helped her wring the towels and sheets and hang them on the lines behind the kitchen to dry. She was amazed that everything had to be ironed to prevent mango worm larvae from hatching and burrowing into our skin. If that happened, an itchy swelling formed with a black dot in the center. When the swelling was covered with Vaseline and squeezed, a small white larvae popped out. The first time this happened to me, and I extracted one on my waist, the kids exclaimed, "Mom! Honestly!" "If I hadn't squeezed it," I told them, "it would have grown into a maggot and after five days hatched into a fly that would bother you or someone else!"

Life was only secure and predictable for Pelesia within the confines of our five acres on the hill. She seldom ventured off the compound except on Sunday mornings, when she attended the service at the Catholic church just down the hill from our compound. Eventually she made a few friends and, much to our consternation, had a relationship with Joseph, Cole's driver. She suffered an ectopic pregnancy, which put her in Nsambya Hospital for a number of days. If we had hired a Ugandan to work for us, there would have been many friends and relatives coming and going from the village. Just walking onto Mbuya compound would have raised expectations, what with the containers full of supplies, vehicles, fuel drums, workshop equipment and packing boxes for shipment.

The man we had brought from Kenya to work for us had been tempted by what was in the containers. Unlike Pelesia, Moses Wanjala, who was from Wabuye on the main road from Malabar to Eldoret, where most of the traffic passed between the two countries, walked to the local market for supplies and made friends with the numerous UNICEF drivers (with whom he caught rides home to Kenya), mechanics and helpers who came and went from our compound. Moses took the brunt of the blame when he and one of the UNICEF drivers who lived on the compound were caught stealing tires from the workshop, and he agreed to leave. We then managed without a gardener for the remaining years we stayed in Kampala. I planted and tended the garden, and Cole mowed the lawn with our electric mower, usually on Sunday afternoons.

House and garden help had been very much a part of my upbringing in India and Cole's and my life together. Before I could walk, Mariamma carried me on her hip. Mary did her best to please me in Bangalore and insisted on spending the night with me when I was harassed by intruders. Just after we were married, Sebastian prepared our morning tea and minded and milked the buffalo in Mandya. Teresa in Bangladesh was efficient and quick. Veronica quietly went about her tasks. Akpan Udo helped us in the house in Nigeria. Gentle and accommodating Antonniamma worked for us in Madras (now Chennai); lighthearted Zenovitch helped us in Ethiopia; Halima with a lovely smile in Somalia; Pushpa cooked sticky rice and Garo curry (but unfortunately also had sticky fingers) in Dhaka; Fatuma in the Sudan was in a category all her own; Anna, Esther, Penina—who I was convinced was an angel—all of them were familiar with our habits and shared our family's concerns. All were friends, with their own familial demands and therefore inevitable financial requests. When we left the countries we lived in, we made an effort to leave our workers with a plot of land, a skill or a sewing machine.

Comfort, Companions and Health Care

The cats we had brought from Kenya were wary of Tiger, our great shaggy dog. Tiger had been shot at close range by an army officer living near Philip Kosana, the UNICEF program officer from South Africa, who unfortunately left the dog behind to fend for itself when he left the country. Cole found the dog in agony and took it to a vet to have the wounds stitched up. They eventually healed. Forever grateful for the security of our chain-link-fenced backyard, though he cowered and whined and came into the kitchen whenever gunshots sounded nearby, old Tiger was well-behaved and never chased the chickens, she-goat, bushbuck and duiker who shared the enclosed space with him. Our younger, more spirited Dalmatian lived in the front garden, and the two dogs would often sleep back to back, separated only by the fencing. The cats, of course, had the run of the house, and usually slept with one of the children. We made certain all our cats and dogs were immunized against rabies, as I had undergone fourteen injections of the antidote around my navel when my cat died of rabies in India (Jhorna and I repeated the painful jabs in Ethiopia, after the puppy our friends had picked up off the street died of the same fatal disease).

Our American Indian daughter, Jessica, had an inherent love of horses. This was something we first noticed when we traveled to Darjeeling on a one-week break from the oppressive humidity and heat and congestion of Dhaka and she begged us for a horse ride. We later purchased a horse in Ethiopia, and the gardener took Jessica and our other two children for rides down the ravine and through the eucalyptus groves several times a week. Although the Hunts had kept polo ponies on the Mbuya property we now lived on, the stables had been converted into quarters for workers and UNICEF drivers. We had been advised that a horse would require a great deal of veterinary attention and vaccines. As medicines were only available in Kenya, we found a spirited Karamojong donkey that grazed on the hill and kicked its heels when Jessica approached it to ride.

Soon after we got the donkey, I heard the back door slam and Jessica burst through the door holding the back of her neck where Punda, the donkey, had bitten her. Although she insisted the bite wasn't deep enough to cause alarm, I washed it with very hot water and soap, apprehensive about the medical care we had access to in Uganda. The country had no geo-political importance in the days of the cold war and Idi Amin's rule had driven all the regional offices from the country and reduced the UNICEF country office staff's presence to skeletal levels. Few were applying for vacant positions and dependants were not allowed. As a result, there was no UN clinic, doctor or nurse. However, Ray Gibbons, an elderly British doctor, ran a small practice at the side of the British High Commission. He did his own lab work and kept records in his file cabinet. Because his clientele was largely expatriate, he knew us all by name.

The probability of us getting malaria in a country where the disease was endemic was, of course, a concern. Fortunately, we lived on a hill apart from heavily populated areas where anopheles mosquitoes spread infection. Still, as a prophylactic against malaria, we slept under mosquito nets and swallowed bitter little Paludrine pills once a week; something I was convinced caused the hum in my ears. It was quite remarkable that none of us had malaria during the six years we lived in Uganda, although many of our friends were affected by the disease. (Years later, Jessica was in a coma for twenty-eight days from cerebral malaria and I was hospitalized with my own bout with the disease.)

God forbid that we would have an emergency, as such cases would have to be flown to Nairobi. Mulago, and what had been one of Africa's best hospitals in the early 1970s, was run-down, with demoralized staff who struggled with blocked leaking drains, water shortages, infestations of rats that could be seen running through the wards and a scarcity of drugs. A few dedicated expatriates taught and practiced at Mulago, including Dr. Goodgame, a Baptist missionary and the Carswells, a husband and wife team of doctors who worked in Uganda throughout the Amin era and into the mid-1980s. Although many dedicated Ugandan doctors struggled to provide medical services, eighty percent of Uganda's specialized doctors fled the country during Idi Amin's reign of terror or in the years immediately following, when salaries fell from $500 a month to $50 as devaluation ravaged the Ugandan shilling. "Uganda's health infrastructure, during the 1960s, was a model for developing countries," Cole wrote in a report dated September 1981. "But in the 1980s the war of liberation caused the total collapse of the country's health facilities."

If hospitalization was necessary, the Catholic hospital at Nsambya provided the best medical care, best doctors, the most medicines and supplies and was therefore where Dr. Gibbons referred his patients to. Our youngest daughter, Jhorna, periodically had convulsions, which Dr. Gibbons diagnosed as caused by temperature or fevers. We then took her to Nairobi to Dr. Colin Forbes, a Canadian pediatrician who had worked at Mulago in the late 1960s, along with dozens of young expatriates caught up with the potential of Uganda after independence and the end of colonial rule. Forbes, who carefully examined Jhorna, noticed the port-wine stain over her right forehead and eye and suggested the convulsions might well be caused by Sturge-Weber syndrome. Cat scans and MRI diagnostic machines were not available even in Nairobi at the time, so we had her examined in the U.S. a couple of months later when we were on home leave. They confirmed Sturge-Weber, and prescribed a prophylactic of phenobarbitol, which, for the time being, controlled her convulsions.

The chicks we purchased in Nairobi grew to be forty healthy hens we penned up for eggs in the backyard. The birds had plenty of space to run and preen in the sunshine, and their noisy cackling became a constant background noise. The children helped feed the chickens, and in time they produced an abundance of eggs. But organizing, buying and transporting the vaccines for the layers was a problem because, like commodities, medicine and poultry meal were not available in Uganda and had to be imported from Kenya. In time, the chickens became the main ingredient for curries and soups. Later, we brought home a handsome rooster, intended for the pot. But our children became accustomed to his strutting and he was allowed to roam the yard with his hen and whatever chicks they raised.

One day, Christopher, a craftsman who sometimes worked for us, appeared at our gate with a small she-goat under his arm as a gift of appreciation. Also a small-time contractor, Christopher gradually built up his business, until he managed to buy an old broken-down pickup. The vehicle was stolen soon after his proud acquisition by envious army men, who then imprisoned Christopher at the army barracks. Araya Assefa, who worked with UNICEF, got word of Christopher's imprisonment and told us the story. We provided money for him to be released. The following week, Christopher appeared at Mbuya, a shadow of the person who had gone into prison, with the goat. Thereafter, Bleat kept us amused with her antics, but was no match for our delicate little duiker, Ibitz, who pranced circles around her.

Aware that we were fond of animals, one afternoon the police guards who stood at the gate had escorted a very young boy carrying a live animal to our back door. "I found it in the bush," the little lad told me, opening his basket to show me an exquisite little bushbuck that looked up at me with large liquid eyes. An unlikely tale, I thought, as poaching in the country was common. There were no wild animals grazing in the countryside along the roads in Uganda, only people, *shambas* and domestic livestock. Most likely the little thing's mother had been snared as part of the bush meat trade.

Although it would encourage the trade of wild animals, I paid for the little creature, which thereafter was devotedly mine. Noey's delicate hooves clicked staccato sounds on the wooden floor when she followed me about the house in search of the bottle of milk I fed her several times a day. At night, she lay quietly in our clothes closet, until she began chewing on Cole's neckties, demonstrating that she had grown enough to sleep outside in the shed. Sheer pleasure was to watch Noey bound through the sprinkler and play with the spray that cast rainbows across her beautiful body.

One warm evening, our home and hillside accommodated the entire student body and staff of Lincoln School. No one even gave thought to driving home after dusk. Full of wonder, we marveled at the tail of Halley's Comet and the moons of Saturn through a powerful telescope Virginia Bobersky (who the children said had taken on head "mattress" duties) had found in the Lincoln School attic—most likely hidden there by the owners when Idi Amin ordered the Asians to leave the country. In time, the skygazers crawled into tents or stretched out on the grass to sleep. The adults in attendance—among them my mother, who had seen Halley's Comet in 1910 as a seven-year-old in India—monitored the stars, planets and the excited children, who huddled in groups around the bonfire under the bombax tree where we had earlier prepared something to eat.

Mother spent time with us wherever we lived, and eventually wrote a book entitled *Sepia Prints* about our lives together. Well into her eighties when she visited us in Uganda, one afternoon she entertained an eccentric Englishman who tipped his hat when he strode up the steps of the house and introduced himself as Lawrence Ascot. Always the gracious hostess, Mother invited Lawrence in for a cup of tea and I later found them conversing easily about his travels to various African countries.

"He was such a gentleman," Mother reported, (almost) ignoring the fact that Lawrence's shirt had been unbuttoned to his navel. Having fled Rhodesia when the country became Zimbabwe under Robert Mugabe, he now ran a little workshop in the city center where he bought pre -World War II motorcycles and cars and exported them to the U.K., where he resold them to collectors at huge profits. Lawrence had come to see Cole that afternoon because they shared an interest in old cars and motorcycles and he had admired the 1956 MGA Cole had purchased in Kenya and driven to Uganda.

Sometime later, Cole found Lawrence dismantling a large military helicopter in his workshop and putting the major components into a steel container that was consigned to the U.K. as scrap. But shortly after Museveni came to power, his army approached the British High Commission, who asked Lawrence to return the helicopter. At other times, he pursued "business opportunities" or did security work for various NGOs and UN agencies. At one point, Cole asked him to fabricate the metal grills for the UNICEF office. Soon after, a suspicious banker presented Cole with a check for the amount of $173,000 that Lawrence Ascot (who had left the country) had forged on the UNICEF account. Reporting the matter to UNICEF headquarters, Cole was told the head office had turned the case over to Interpol, with the result that Lawrence spent time in a U.K. jail. Much later, we read in a British magazine that Lawrence had again been arrested, this time for selling a bogus HIV Aids cure (in reality Congolese water) in Europe, having abandoned one of his Ugandan families. By all accounts, Lawrence was one of those practiced con artists who took advantage of situations and countries in chaos. They were drawn to the lawless opportunities of Uganda in the early 1980s, taking advantage of their white skin to corrupt their way around a country with no retired settlers or legitimate investors, and certainly no tourists!

The wealthy and committed Oxfam Field Director, who worked with the famine in Karamoja, was a different sort of expatriate. He was attracted to working with an unusual situation in Uganda. Acutely aware that the greatest security risk he had was the drive between his office in Kampala and the projects in Karamoja, he purchased and flew a Cessna 182. He also had a helicopter license and was rumored to own a Bell JetRanger that he flew around England, apparently landing it on a school sports ground in Summertown, north Oxford (where Oxfam was headquartered) when he traveled to his interview for the position of Field Director.

Often of an evening, our frequent houseguests joined us in what they said was a "wonderfully normal household" given the insecurity they otherwise felt in a country at war. When the sun disappeared over the hills, we pulled our chairs close to the fireplace and shared about the activities and occurrences of the day around the flickering flames. In years past, when we had worked for the British charity Oxfam, it was expected that anyone working for the organization was welcome to stay with the family of the field director. It was different in the UN system, with many more nationalities, larger paychecks and per diems covering five-star hotel costs—but since the hotels in Uganda were rundown and ill-equipped (if they functioned at all), our visitors stayed with us. Our family routines were in stark contrast to the lifestyle of the UN expatriate staff in Uganda, most of whom were young and single. Many lived with their girlfriends or boyfriends or together in a shared accommodation, especially those on short contracts. Their schedules were centered on work, although in those pre-computer years, there was no easy way to take work home. All-night parties on weekends were the order of the day, as the UN system imposed a curfew, which was broken from time to time.

One sundown, as we sat around the fire, Jhorna removed the

headband that held back her thick black hair and solemnly said, "Our French teacher didn't come to class today." And as she spoke, I stared at the embers that suddenly sparked into flames.

"The other teachers said she was attacked and everything she had was stolen," she continued. Concerned the ongoing stories of violence would affect the children, and recognizing their need to talk about them, I kept the discussion going. "The thought of losing our possessions doesn't affect me as much as . . ." I began.

". . . as people coming with guns," Sanjeeb finished my sentence, drawing his chair closer to the fire.

"Anyway, we could jump behind the fence and hide in the bushes if they come," Jessica suggested—a comment that put a sliver of fear in my heart as I watched the silver moonlight shimmer across the distant waters of Lake Victoria. For not only were we vulnerable, but it was sobering to have our children exposed to theft and violence as if it were a normal occurrence. It also surprised me that they had developed strategies of their own.

That night, after we had closed the doors on a luminous sky crowded with stars, Cole and I found few explanations for the children's queries regarding the occurrences of the day, but continued with what we knew to be helpful routines. A few stories and Bible verses later, we tucked them under their mosquito nets and Cole's voice rang out with, "Michael rowed the boat ashore . . . chills the body but not the soul."

As was our custom, before getting into bed, I checked the "mailbox" we kept on the top of the clothes cupboard to see whether the children had left us any messages. Although Sanjeeb was blessed with special artistic talent, all three children spent much time around the low round table they had in their room with plenty of paper, crayons, pencils and paints that allowed them to create art whenever they liked. They often decorated the notes they wrote us with floral designs, stickers, pressed flowers, cutouts and colorful drawings. Sometimes their little grievances or concerns made us laugh. Other endearing little notes offered reassuring words, such as "Life is so good because God is good." One of them: "When there is love, there is kindness, when there is kindness, there is gentleness, when there is gentleness, there is peace, there is joy," made me tear up. At times, a response to their notes was in order, and Cole or I would "post" letters back to the kids, who checked the box regularly.

Our three children had richly enhanced our lives.

Bat Valley

Cole took up the challenge of his new assignment, determined to make a success of the mess he inherited. The office was made up of five rooms in the old Uganda Teacher's Association (UTA) located on Bombo Road in Kampala, near the center of the business district and close to Makerere University. Cole's predecessor, Jim McDowell, a nutritionist, managed the small country office (which had been UNICEF's regional headquarters from 1962 to 1971, before the international staff fled to Nairobi) as best he could under the terrible and trying times of Idi Amin. His staff had consisted of an international program officer, a transport officer and nine Ugandan support staff, before he was abruptly transferred. McDowell then drove the Representative's Volvo saloon to Nairobi, where Cole picked it up and drove it back to Kampala. Philip Kosana was transferred and Mark Pollan left, as his contract was finished.

"I will never forget the day Mr. Dodge took over the office," Joseph told me one day, as he drove me to meet Cole at a reception the British High Commissioner was holding. "That day, *mzee* called a meeting and took all of us around the office and assigned us our positions and where we would sit. After that, he had discussions with each section before he asked us to remove our personal items." Cole collected the keys that locked the security grates and doors to their offices, desk drawers and cabinets, and requested five men to work with him over the weekend. One by one, piles of boxes, stacks of old files and used tires that had accumulated over the years were carted downstairs and thrown into the back of a truck. Cartons of pens, pencils and paper were discovered and set aside to be delivered to the Ministry of Education for use in schools. The stacks of old files were taken to the city council dump, where they were set alight with kerosene. Vehicle parts, boxes of outdated UNICEF calendars, broken chairs and desks that could be repaired were taken to the parking lot and divided into five equal stacks with a number on the top of each pile. Cole then asked the men who had helped him clean the premises to select a number he had written on his fingers. The number they chose matched a pile, which was then theirs. Having cleared everything out of the offices, the men hosed down the rooms from ceiling to floor with a pipe attached to a water pump and swept the water down the stairs onto the parking lot below. Cole divided the largest room into administration, finance, logistics and supplies. Those handling health, water and sanitation and education would use the next three rooms. His own office he set up in the back room with a view of the thousands of bats that hung from the numerous eucalyptus trees that grew in the valley below, which was fittingly referred to as "Bat Valley."

Although Cole had nine UNICEF staff when he arrived in Uganda, the program soon exploded to over three hundred by the middle of 1981, in response to the drought emergency in the Karamoja. Consultants, international reporters, donors such as the Canadian CIDA staff from Nairobi and Ottawa, UNICEF regional office and headquarter staff were curious about such dramatic program expansion, particularly within the context of insecurity, civil war and the breakdown of civil society. A combination of factors influenced Cole's achievements, including the inspiration he got from James P. Grant, UNICEF's executive director, who had just been nominated by President Jimmy Carter to head the organization. Cole also got on well with the determined Swedish academic, Karl Eric Knuttson, who Jim Grant appointed as the regional director in Nairobi. Having led NGO field offices in half a dozen countries, Cole had ambitious goals and simply got on with what needed to be done without following the hierarchy typical of the UN system. He also benefitted from the commitment and enthusiastic capabilities of the professionals he hired. Although Carl Graham had reached the retirement age of sixty, he was given a contract extension as Cole's deputy. He and his wife, Trudy, lived in one of the other two houses on our compound. The Canadian doctors Liz and Don Hillman, who had experience in Kenya, Zambia and Uganda, were in their late fifties when they joined UNICEF as advisors to the health program. Most of the other staff—Brendan Doyle, Peggy Henderson, Sally Fegan, Mark Stirling, Robin Biellik, Rohini D'Silva, Vishwanathan and consultant Francis Mburu—were younger but highly motivated.

Blood Red

Bees swarmed in the afternoon heat, and now and then a sunbird or butterfly rested on the numerous flowers near the verandah. Billowing white clouds contrasted sharply with the blue sky. The view from our Mbuya Hill home included the sheen of Lake Victoria far to the south, the verdant valley with its clusters of homes and other buildings below and Tank Hill just across from us. Built on seven rolling green hills, the city of Kampala to the right appeared peaceful in the distance.

Our children, Jessica, Sanjeeb and Jhorna, off from school for the day, complained in their customary way when I called them in to practice their piano lessons. Later, like squirrels, they resumed their climbing of the flamboyant tree and took turns swinging from a rope tied to a large branch. Pelesia hummed a tune as she collected the laundry from the clothesline.

Although it was mid-week and I normally made bread on Saturdays, I mixed yeast into warm water in a large stainless steel bowl and added salt and oil and a tablespoon of sugar before sifting in enough flour to make four loaves. While I was working, tousled old Tiger, the dog Cole had rescued from a nearby compound after its residents had fled following upheavals, nudged the screen door open and flopped under the kitchen table. He was still dozing an hour or so later when I placed the bread pans into the oven next to the lasagna. It was warm in the kitchen. Straightening my back, I looked out the window and was taken aback at what I saw.

"Kids, come and see!" I exclaimed.

Beautiful as the blood-red sunset was, perhaps it was an omen of what was to come, for within minutes, explosions rocked the stillness of the evening and shattered the calm of our day.

The shortwave radio sputtered to life and I ran to turn up the volume.

"Heavy shelling on Mbuya Hill," the United Nations Security Officer told all stations. "Remain where you are."

I could hear shouts and gunshots above the roar of vehicles racing down the hill from the army barracks across the road. Flares lit up the darkening sky. Machine guns clattered from a number of angles. Fearing the worst, my only thought was flight.

"It's Papa!" Sanjeeb shouted above the tumult.

Taking the telephone receiver in my trembling hand, I held it out for Cole to hear what was happening.

"Is the back gate locked?" he responded helplessly. He was calling from Nairobi, where he'd gone for a donors meeting and to finish his flying course. "Stay as calm as you can," he continued. "If they come for the vehicles"—twenty or so trucks and cars were lined up near the UNICEF workshop just down from our house—"give them the keys without fuss. God be with you. I love you."

Explosions continued to shake the ground as I put down the receiver. Frozen with fear, the children stood nearby. I wrapped my arms around them and we huddled together and asked God for protection. The explosions were deafening. With escape on my mind, I asked the children to get a warm wrap and whatever else they wanted. My fingers fumbled open the desk drawer in which we kept our documents. Tucking our passports into my shirt pocket, I led the children to the back fence of our lot, where we stood near bushes under the flamboyant tree. The night was filled with gun smoke. Then, as suddenly as it had begun, the roar of vehicles, gunfire and explosions stopped. Weak with relief, we sank to the grass to welcome the quiet and gentle breeze that soothed our stretched nerves.

That troubled night, I lay awake long after the children had gone to sleep, listening to the thud of heavy artillery from the direction of the Busega swampland southwest of the city, for I knew that further battle was imminent.

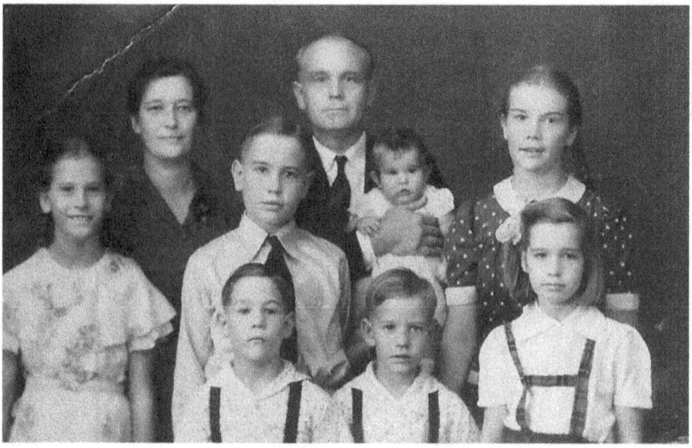

Family photo, 1944. Top row (left to right): Mother, Father holding Marilyn, Esther. Second row: Ruth, John. Front row: David, Paul, Irene.

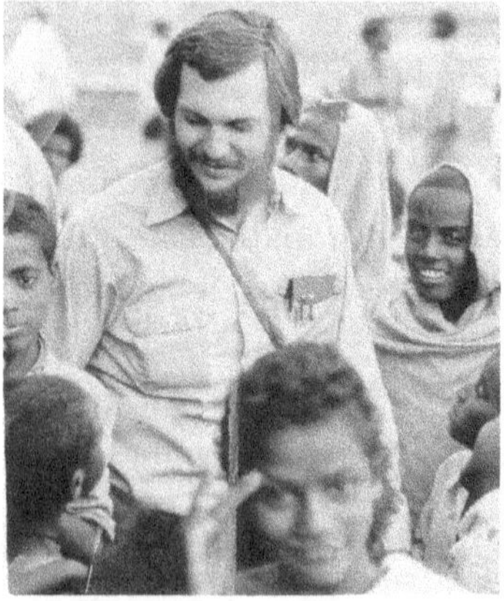

Cole in Tigre, Ethiopia, 1973

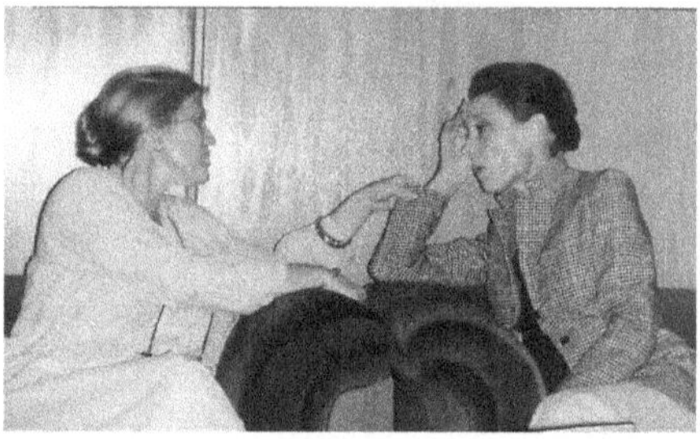

Marilyn and Audrey Hepburn in Bangladesh, 1991

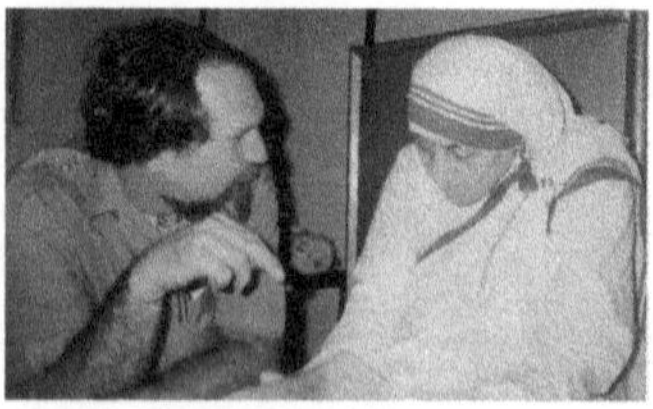

Cole with Mother Theresa

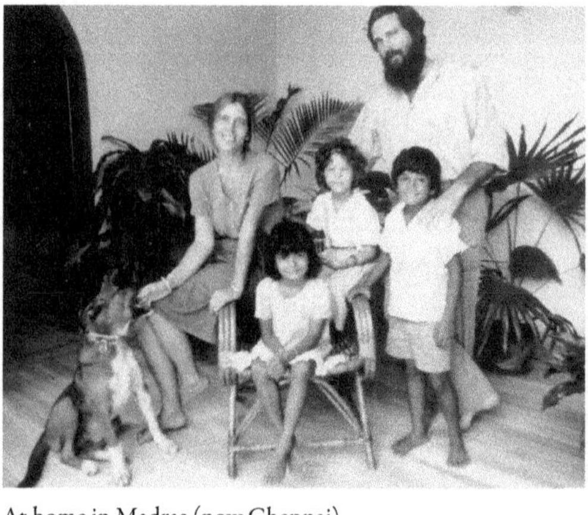

At home in Madras (now Chennai)

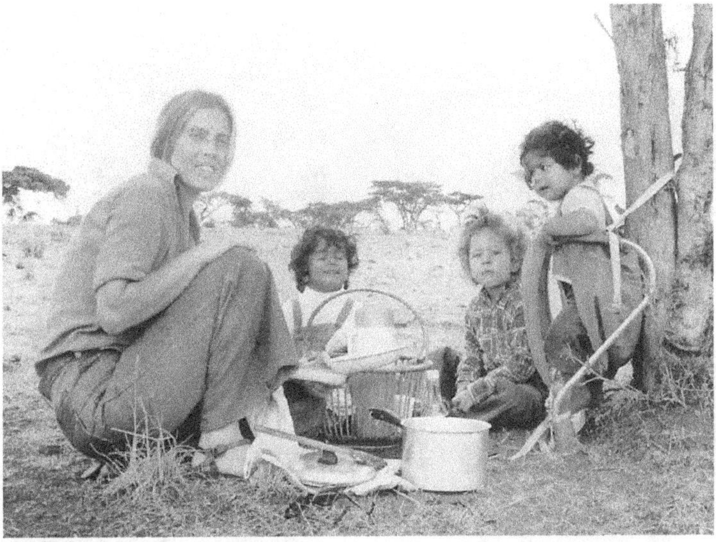

Marilyn, Sanjeeb, Jessica and Jhorna picnicking in Ethiopia, 1976

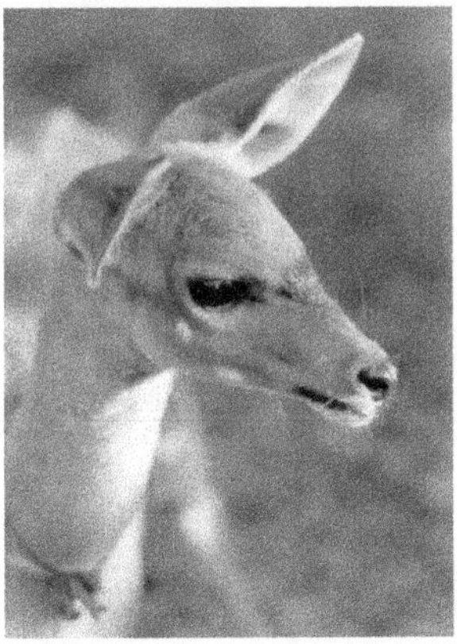

Noey

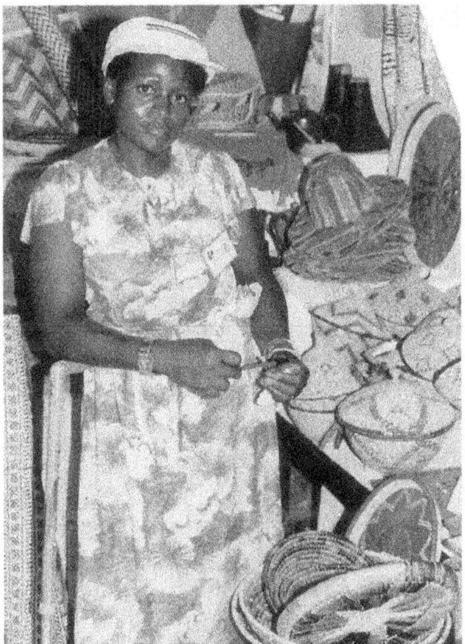

Betty Kinene at Uganda Crafts, 1974

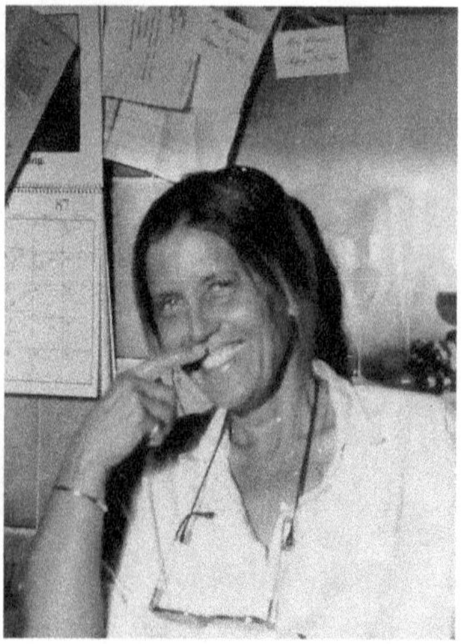

Marilyn in Khartoum

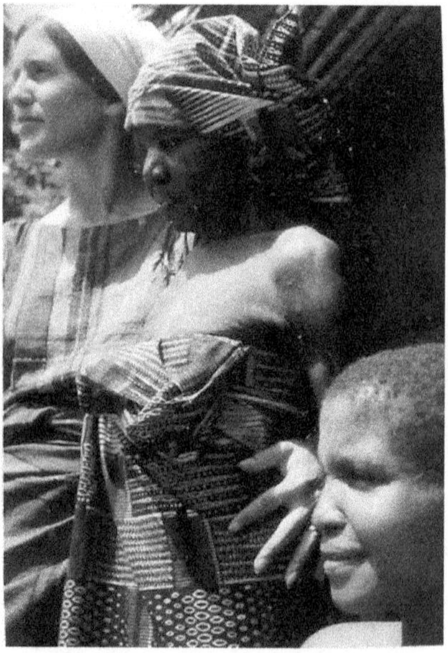

Marilyn and friends in Ikot Usen, Nigeria, 1970

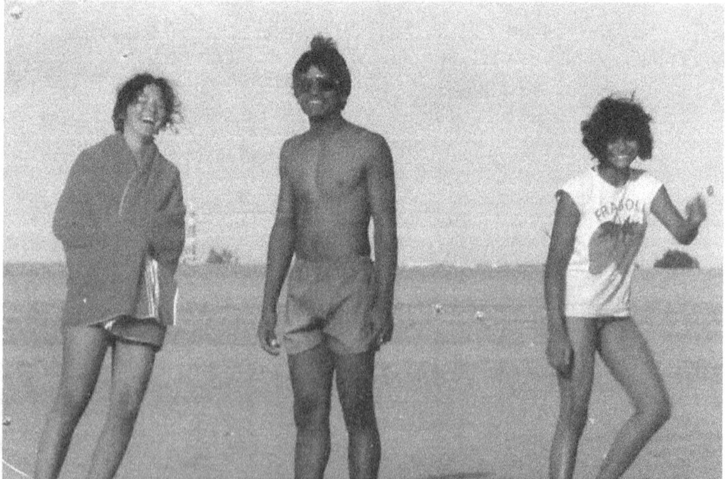

Jessica, Sanjeeb and Jhorna on the Nile

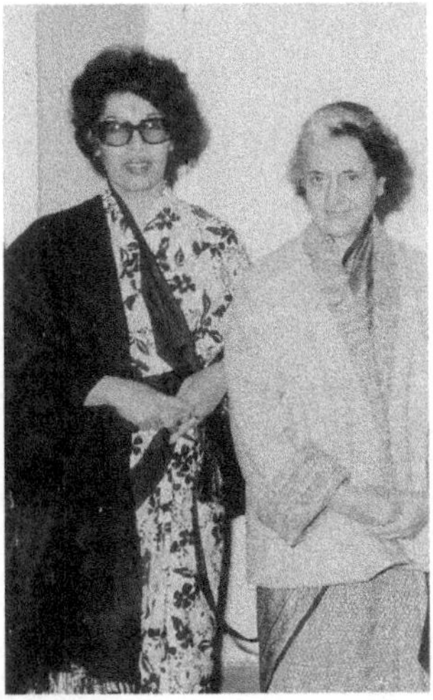

Kuku Barua with Indira Gandhi

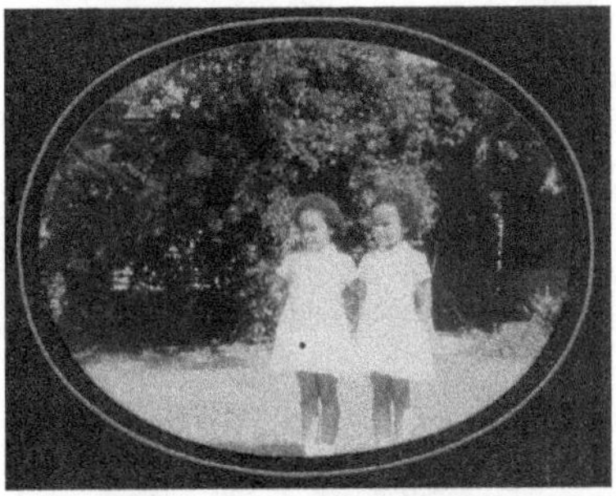

Penny and Pat as children

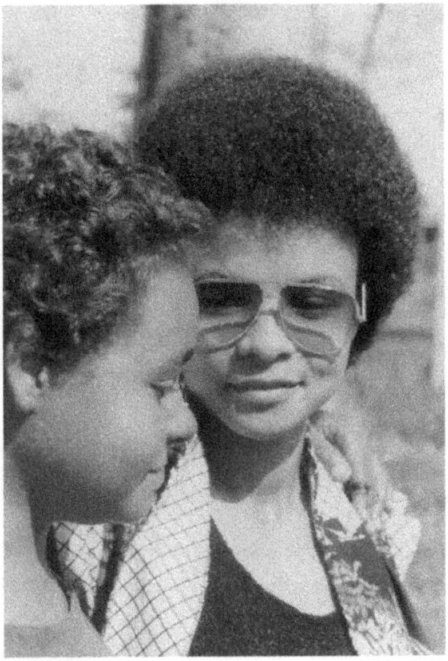

Pat and Penny

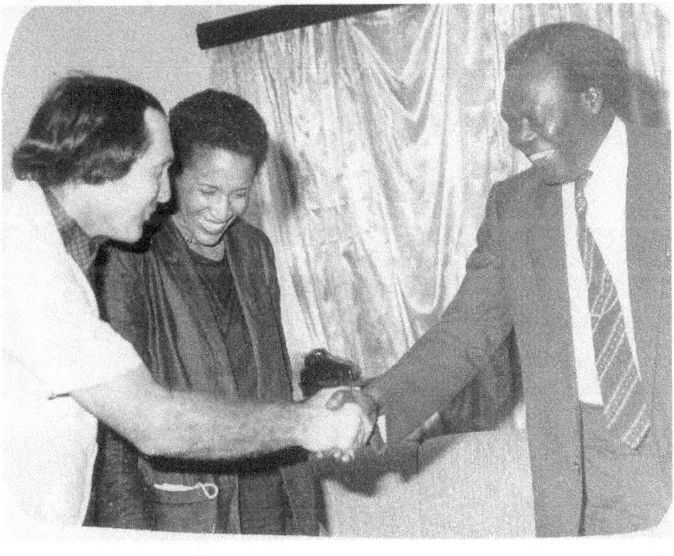

Left: Cole with Milton Obote, 1974

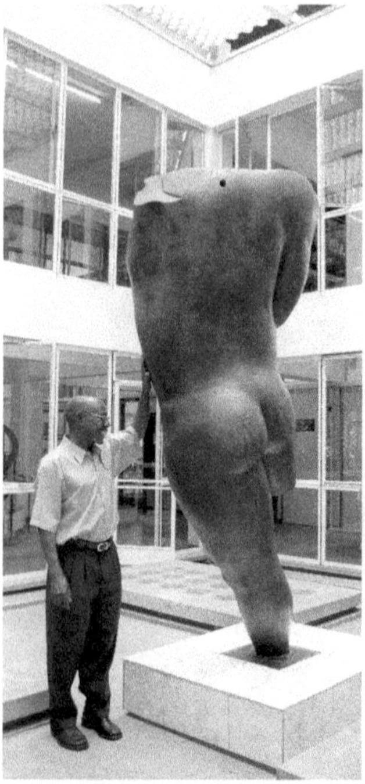

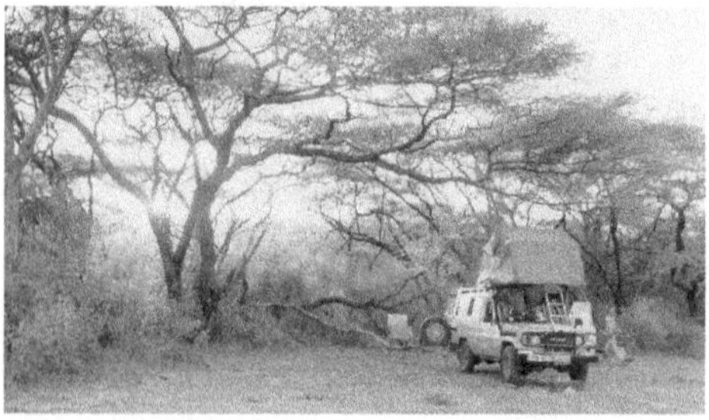

Setting up camp

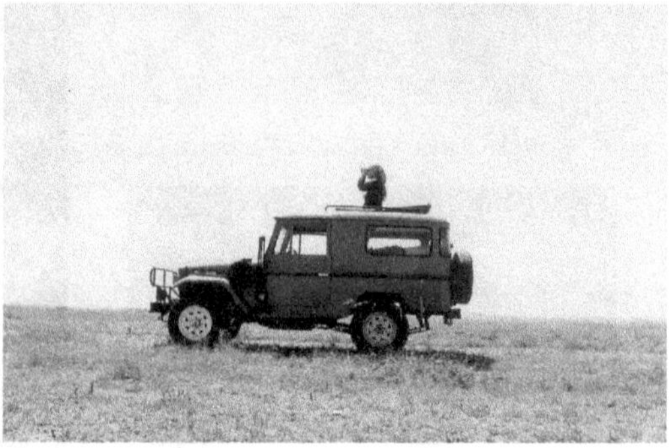

Francis Nnaggenda with "War Victim"

On safari

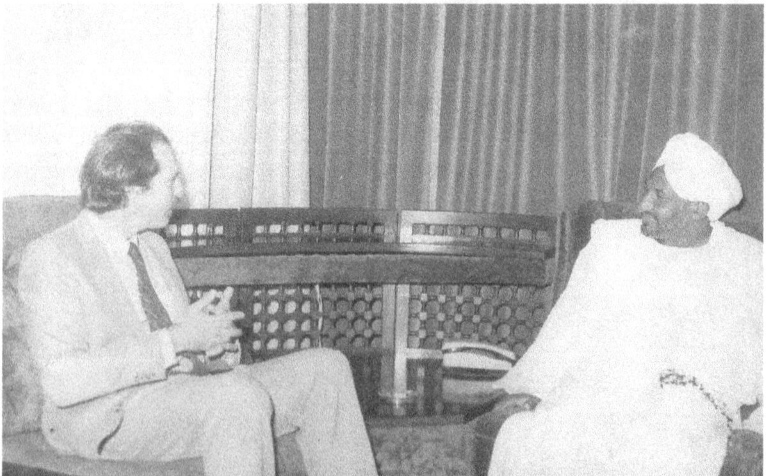

Cole with Sadiq al-Mahdi, 1987

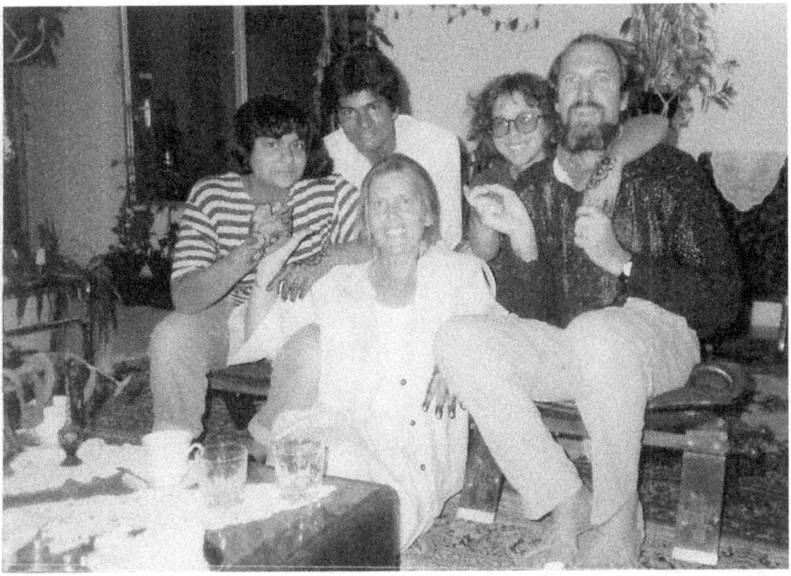

Left to right.: Jhorna, Sanjeeb, Marilyn, Jessica, Cole

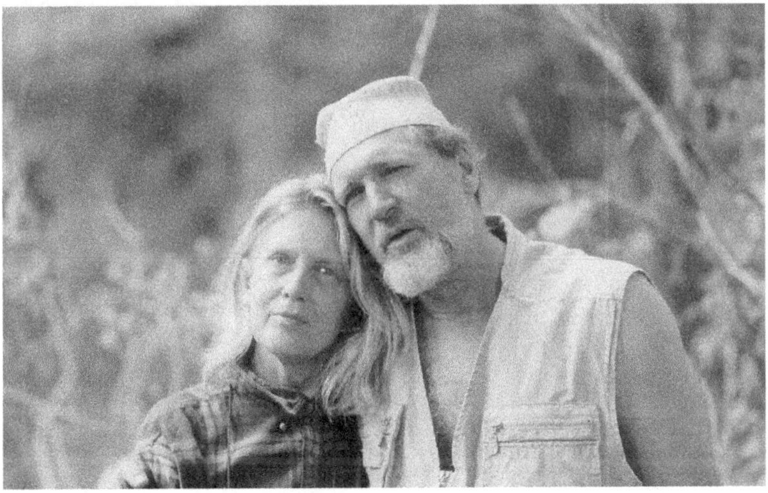

Marilyn and Cole, 2010

Chapter XII
Famine

"Blame is laid on nature, which persists, year after year, in following its fixed pattern. But man is responsible . . ."

—Mort Rosenblum and D. Williamson, *Squandering Eden*

The Karamoja

CONSISTING OF MILE after mile of untamed grass plains, the Karamoja in Uganda is bordered on the east by the steep escarpment of the mighty Rift Valley that descends two thousand feet into the Turkana District of Kenya, to the west by the Labwor Hills and to the north by the Sudan, which was engulfed in civil war. Ethiopians were fighting Somalis with arms supplied by outside powers that found their way to the Karamojong, who bartered their guns for cattle and looted Amin's armories. Major skirmishes also occurred between the Dodoth in the north, the Jie in the center and the Bokora, Pian and Matheniko in the south. During one such conflict, Médecins Sans Frontières personnel mistakenly chose to disregard the advice given them by the Catholic fathers and left the wounded in the care of the medical staff at the hospital in Moroto. Later that evening, long dark shadows crept past windows left open to the twilight air. Armed Jie entered the ward and speared the enemies of their clan while they lay in their hospital beds.

When Tanzanian forces liberated Uganda of Idi Amin's tyrannical rule in 1979, the Matheniko raided Moroto's well-stocked armory in the Karamoja. Over a three-day period, women and even children helped their men carry the rocket-propelled grenade launchers, AK-47s and landmines to their homesteads, where they hid the loot. Obote then sent his army to quell the raiders. After this, raiding, always a part of the Karamojong tradition, took a lethal turn. "The formation of raids, that of surrounding, was known more than a century ago," Elizabeth Marshall Thomas writes in *Warrior Herdsmen*.

> "(The) advance is stealthy; no one speaks . . . When the leader decides the time is right, the columns burst from cover, the men, shouting the names of their oxen, form a huge ring which tightens gradually, drawing cattle and people close to the abuthia. The raiding party then reverses itself for the escape. This time the columns move side by side, forking open only at the forward tip, where the captured cattle run."

At the same time, the balance of power between the local and cross-border Kenyan rival groups plagued the Karamoja. The Turkana used firearms to raid and eliminate whole settlements of the Karamojong, who offered little resistance with their spears.

Spears became obsolete and violence escalated out of control, with dozens killed, whole families wiped out in a single raid and whole clans left destitute when their livestock was taken at gunpoint.

Drought

Before the children and I arrived in the country of Uganda, drought stalked the land of the Karamojong. The hard earth cracked. Tanks dried up. Women were no longer able to reach water with the holes they dug in sandy riverbeds. Herders moved their cattle to where they could find pasture when dust devils whirled over the barren soil. The hot wind continued its relentless sweep over the wide plains and left the grass brown and brittle, the earth thirsty and gray.

Watching the landscape wither and dry, the Catholic fathers who lived among the Karamojong alerted donors to the potential emergency, but Uganda was such a broken land, the UN system was inadequate and disorganized and the NGOs were competitive and uncoordinated. The reaction was far too slow. By the time Cole arrived in Uganda in January of 1981, Father Angus Finucan, who ran Concern in Namalu, two or three Irish nurses with GOAL, Lutheran World

Federation, SCF, MSF, Oxfam and a host of church organizations had begun to respond to the famine with emergency supplies and relief teams. Although UNICEF had much more emergency relief capacity and experience than the United Nations Development Program (UNDP), Melissa Wells, who had served as an ambassador and a career diplomat in the United States foreign service, was assigned as the Secretary General's Special Representative to Uganda under UNDP to set up an inter-agency coordination committee. This resulted in the two UN giants competing for turf more often than they cooperated.

"Go ahead and proceed as though you have the money," James Grant advised Cole, who, together with Karl Eric Knuttson, put together a three-and-a-half million dollar budget to hire staff, order vehicles and "service" the NGOs' child-feeding operations in over two dozen feeding centers for malnourished children. Emergency staff was recruited, trucks and Land Rovers ordered, contracts signed, coordination meetings called and plans laid out in such dramatic ways that UNICEF wrested the role of lead agency from UNHCR and gained donor recognition, which resulted in funding. Within four months, the staff had risen to 200, and by midyear the payroll had 360 employees.

Three intense months later, personnel, including expatriate staff, local doctors and nurses, many drivers, mechanics, warehouse workers and nutrition scouts distributed 2500 tons of maize, oil and skim milk powder (provided by the World Food Program) to twenty-five child-feeding centers and organized the immunization of children against the five principle diseases. They used Soroti's abandoned Barclay's Bank, the Shell petrol station and the old meat-packing plant for the warehouse and logistics. "It is as if UNICEF owns the town," the people said.

However, not everyone supported this massive food-aid effort. On 15 May 1981, a letter addressed to Cole from the Free Uganda Movement warned:

Gentlemen,

We introduce us herewith as members of the Free Uganda Movement. As we have mentioned before on some occasions, our goal is the ous-

ture of the present government of the drunkard Obote. We appreciate the goodwill UNICEF is giving to our brothers in Karamoja but please understand that your service at the present time is a sort of backup for this government.

We will need your help in the future of Uganda but we request you in a very serious way to withdraw your entire personnel from this region in order to allow us to pressure this government into resignation.

Should you fail to oblige we will be forced to take some of your staff hostage til you recognize that this is no drill but a serious matter concerning the liberation of our country.

FUM, Executive Committee, Soroti

The information was noted, but Cole continued the program despite personnel conflicts, harsh working conditions and extreme insecurity that took the lives of six UNICEF drivers and their assistants.

Bands of young warriors regularly roamed the countryside and set up roadblocks on largely deserted roads to rob the UN and NGO personnel. At times, despairing at the arrival of so many relief agencies and their vehicles, the priest who ran the relief operations for the Catholics claimed they had never been robbed, because they had worked with the people in the neglected region for so many years. Then, one day in 1982, Father Mario and four nuns were stopped at gunpoint at Irriri and stripped of their clothing. Father Mario then drove his Land Rover to the mission compound and halted at the junction in Moroto—with one road leading to the father's quarters, the other to the sisters—until eventually someone was seen delivering an armload of bed sheets to the missionaries. Some days later, a long-limbed Jie tribesman was spotted walking stark naked but for the white clerical collar encircling his neck. Shortly thereafter, Father Mario, an anthropologist by training, suffered a breakdown and left the country.

When the severity of the famine, which had taken the lives of twenty-five thousand people, showed signs of abating, many international agencies and their staff withdrew. It then became almost impossible to interest people from outside the Karamoja to return, due to the region's collapsed markets, lack of security, poor roads and scarce bus and postal services. In order for the Ministry of Health to continue operating twenty-five nutrition centers, the Karamojong, with the lowest literacy rates in the country (5% for women and 17% for men) were recruited and trained with UNICEF funding to fill personnel positions in nutrition surveillance, immunization and other aspects of basic health care. Gradually, experience became the workers' best teacher and local capacity emerged.

Once again, the Karamojong turned their anxious faces toward cloudless skies when, in addition to another drought, conflicts between the Karamojong and the Turkana became so intense that a joint military operation by the armies of Kenya and Uganda was mounted in an attempt to disarm the raiders and restore order. Unable to hoe their fields, as many as thirty thousand people who had settled in the fertile plain surrounding Namalu in the southern part of the Karamoja were forced to flee to the nearby foothills of Mount Elgon. Fortunately, this time twenty-five feeding centers were functioning near major settlements jointly supervised by the Catholic and Protestant missions. Mothers brought their children to have their nutritional status assessed by weighing the child on a spring balance to compare the child's weight to the child's height. Wherever possible, food was distributed as payment for public work such as road repair and the construction of schools and clinics or clearing new ground for cultivation.

TEMO

Eventually the severe situation in the Karamoja improved and the UNICEF program transitioned from relief to rehabilitation. The audit UNICEF then commissioned found that over half of all supplies (drugs, spare parts, vaccines, refrigerators, sterilizers and medical kits) were stolen from the ministry offices, medical stores and repair shops in Entebbe before they reached the rural health centers. Faced with overwhelming insecurity and with leaky government warehouses and the breakdown of medical equipment and vehicles, UNICEF and the Ministry of Health set up a Transport Equipment Maintenance Organization (TEMO). The Ministry offered UNICEF an abandoned Asian property with high walls opposite the Lugogo Stadium on Jinja Road, which was ideally located, from a security and management point of view, to set up TEMO.

Araya Assefa was then transferred from the Karamoja, where he had successfully managed the fleet of trucks carrying food to feeding centers and distribution points, to manage TEMO, train Ministry staff and set up the largest health program in a decade. But, despite precautionary measures, which included armed guards and an electric fence, thugs chipped their way through TEMO's concrete wall and stole everything. It was obviously a well-organized robbery, which probably involved criminal elements of the police or army, since trucks were used to carry the loot. Despite these setbacks and the regional office based in Nairobi not really recognizing Uganda as a priority country, the Kampala UNICEF office was on a roll, with support from interested donors and energetic staff largely allowed to do whatever they could, provided they raised the needed resources.

Soon after organizing TEMO, UNICEF purchased the five-acre Mbuya property we lived on for $200,000, in what turned out to be a wise investment. Although the three houses on the compound were run down, they were soon renovated and the chain-link fence around the complex was repaired. Some of TEMO's supplies were then transferred to the compound to be stored in the heavy metal containers that could not be broken into and workspace and parking areas were laid out near the workshop for the UN vehicles. The workshop was located just fifty meters below our home, which I feared would increase our exposure and was sure bait for looters. Although it was a risk to store spare parts, fuel reserve and two dozen vehicles used in UNICEF's health, water and sanitation programs on the property, there really was no alternative.

One morning, Isaac from Tororo, who had been hired to manage the workshop, having also worked with the famine in the Karamoja, dropped to the dirt when a bullet smashed the rear window of the truck he was working on. Fearing the worst, he kept his head low until he peered over the truck to see one of the six policemen who guarded the premises walk toward him. The guard greeted Isaac with a sheepish grin and an apology and admitted that he had been careless with his gun. Although Isaac chuckled when he told me of the incident, I was alarmed at the combination of bullets and close calls, and prayed none of us would be in the wrong place at the wrong time. From the start, I thought it an imposition to have twenty-four-hour security guards patrol the gate and fences of the compound, but of course it was necessary. The guards were part of the local police force organized by UNDP and, unlike their Ugandan colleagues, enjoyed extra bonuses, such as monthly commodities purchased in Kenya and a midday meal brought to them by mobile meals-on-wheels.

The lawless nature and grinding poverty of Uganda created circumstances where deception and ingenuity thrived. Regrettably, misappropriation was common. One day, one of the Suzukis was hijacked when it was taken for a test drive. (After that incident, UNICEF increased internal controls.) The same evening, Cole heard a commotion coming from the workshop and walked down the hill to inquire about the problem. Unable to conceal his wet crotch, Dominic, a tall Acholi who worked for UNICEF, told Cole he had urinated his pants when, as he claimed, "They hijacked the car I was driving." The Suzuki was later recovered by the UN police unit. Cole had Dominic investigated, which revealed that the "soiled pants" episode was a tale concocted to cover his tracks for colluding with the thug who actually paid him for the Suzuki. Unfortunately, Dominic did not change his ways, and when he was later caught stealing tires from the vehicle spares store, he was encouraged to resign. It was decided not to turn him over to the police, as the treatment in the jails was simply too brutal. We later heard Dominic found a job working as a UNICEF driver in Juba, in southern Sudan, where he died of Aids.

Ombachi

When Idi Amin's army retreated, they pilfered the rural health centers, clinics and regional hospitals in the northern part of Uganda. What they left behind was taken by the liberation and Tanzanian forces.

Frightened doctors, nurses and medical assistants went into hiding and, of course, were hesitant to return, fearing further attacks.

On 23 June 1981, the Catholic mission three miles north of Arua at Ombachi (the headquarters of the Verona Fathers' West Nile schools, health centers, churches and hospitals) was attacked. Eighty-two people were massacred and fifty-three injured in the mission compound. The massacre at Ombachi shocked the Catholics and caused NGO and UN agency personnel, as well as long-term missionaries, to be evacuated south to Nebbi and Angal, a predominantly Alur tribal area close to Lake Albert and the Albert Nile. Thousands of civilians fled across the borders to the Congo and the Sudan, as well as more than a few UNLA soldiers. Arua was virtually abandoned.

Two days after the Ombachi bloodbath, Cole drove a Land Rover packed with emergency medical supplies toward Pakwatch Bridge, which spans the Albert Nile at the mouth of Lake Albert in northwest Uganda. He was alone. Fearing the probable encounter with Amin's Kakwa soldiers, UNICEF drivers chose not to accompany him to the West Nile.

Soldiers blocked the road ahead. There was no chance to turn around and retreat. Making sure the doors of the Land Rover were locked, Cole halted on the one-way bridge, very much aware that soldiers at remote roadblocks were known to bully and sometimes kill their victims, in order to loot. Leaving the hot engine running, he watched nervously as a man in a faded uniform approached the car. The soldier brandished his gun and ordered Cole to get out of the car. Cole refused. The soldier then ordered his subordinates to force open the doors of the Land Rover. Alarmed, Cole gunned the engine and turned up the volume of the two-way radio to assure the soldier (and himself) that he was in contact with Kampala. Not at all convinced that information was getting through, he shouted into the mouthpiece and welcomed the staccato exchange with the UN security chief. The soldier standing near the window of the Land Rover was restless and shuffled gravel under his feet, until suddenly he pulled a pistol from his camouflage trousers and held it to Cole's head.

"The hard metal of the gun felt oddly unthreatening," Cole told me later, "though I remember the soldier had beads of sweat lining his upper lip and I could feel sweat running down my neck."

Curious about the stalemate, the soldiers who had been loitering in the shade under the bridge sauntered over to join their commander. Their conduct pregnant with ill will, they cupped their hands against the dusty windows and peered at the boxes packed to the roof of the Land Rover.

Fearful, and mindful of how vulnerable he was, Cole slowly reached for the radio mouthpiece and again spoke loudly into it, hoping the UN security officer in Kampala sensed his predicament. He did, and asked Cole to give the mouthpiece to the soldier, who was told to look at the license plate of the vehicle Cole was driving.

Fortunately, the man took note of the Corps Diplomatic 96 license number and Cole was immensely relieved to hear him say, "Let him pass."

Cole then eased the Land Rover into gear before the soldier could change his mind and sped toward the distant dark mountains. After turning off the main road, he diverted through Nebbie, where there was a mission hospital, and on to Gole on the Zaire border. At the huge Catholic cathedral in Warr, he stopped and talked with the elderly Italian bishop, who gave Cole his blessing and said, "God willing, you will reach Ombachi." But before Cole reached Arua, weary foot travelers told him the mission had been evacuated. Nonetheless, Cole drove on to Ombachi and confirmed what they had said. Horrified at the sight of blood on the mission floor (the corpses had been hurriedly buried), Cole decided to drive back to Warr, where he unloaded the medical supplies with the Italian bishop who, despite having witnessed the recent massacre, had chosen to stay. Offering the old man words of encouragement, Cole began the tiring journey back to Kampala.

Ongoing disturbances, in combination with a long period without rain, displaced 180,000 people in West Nile province. Identifying the most affected areas with a survey conducted together with the Institute of Public Health at Makerere University, UNICEF set up supplemental and therapeutic feeding centers for the malnourished, as well as curative clinics and immunization programs, with particular priority given to small twenty-bed wards and maternity centers.

Four thousand people fled their homes in the market town of Adjumani located on the eastern side of the Nile in the extreme north of Uganda. Overwhelmed with the problem of feeding and caring for so many people, the Catholic mission provided the best facilities they could. But, convinced a health disaster was about to occur, the UN team organized men to dig latrines some distance from the camp while pits were dug within the compound for the small children, most of whom exhibited severe signs of malnutrition.

Around this time, Jim Mayrides, the UNICEF desk officer from New York, requested that Cole charter a plane to visit the disturbed area of the West Nile. The Aviation Sans Frontières pilot ("Pilots Without Borders," whose organization's first assignment was in Uganda), who had flown commercial airlines for Air France, told Cole he had not flown a Genet single engine for thirty years. "Since training," he said, as he held back the joystick and gave the plane full power. Guiding the swaying, shuddering post-World War II airplane down the runway until it was airborne, he flew below the clouds by visual flight rules. Scanning the sectional map, the pilot set their course for Arua, toward the Nile River and Lake Albert. Unhappy with the low clouds to the west, he flew to the northeast, hoping to skirt the cloudbank. Lower and lower they flew, until the clearance between the clouds and ground was only about five hundred feet. Deeply concerned, the pilot turned the plane 180 degrees, only to find the clouds had closed the route.

The pilot then circled until, breathing a sigh of relief, he and his passengers could see the River Nile below and, although heavy mist engulfed half of Murchison Falls, they flew a hundred feet above the river toward Lake Albert and followed the north shore to Pakwatch Bridge. By now, the clouds had lifted a little, and they could see railway tracks below, which Cole knew would lead them to Gulu. Gaining elevation, they sighted Gulu, and finally were able to land on a concrete runway built for air force jets and large transport planes.

Realizing they had not filed a flight plan for Gulu, the men were so relieved to be on the ground that they jumped from the plane, oblivious to the soldiers who raced toward the plane with their guns cocked. Jim, however, remained seated, obviously in a state of shock. Anxious about his condition, Cole punched the frequencies on the radio for UNICEF Kampala and requested they contact Lachor Hospital for emergency care. Even the army men hovering about were reassured when, fifteen minutes later, the husband and wife team of doctors, the Cortes, pulled up in an ambulance and gave Jim a shot of adrenalin. Flying at treetop level had affected all the men. When Jim recovered, he simply refused to get back on the plane, and the pilot said he would not fly again that day. Instead, the men revived their spirits by talking about their traumatic experience over a glass of beer, *matoke* and stew at the Acholi Inn in Gulu.

Chapter XIII
Tales and Safaris

In Our Custom

I HAD ONLY HEARD about the devastating famine in the Karamoja and points north in the country of Uganda, for I had not had the opportunity to visit.

Hearing a car door slam early one afternoon, I assumed the children had returned from school. But it was Vicki, a tall, attractive Karamojong woman, who walked up the stairs.

"Am I welcome?" she asked, kissing me on both cheeks. Pleased to see the woman I had met earlier at a reception, I led her to the verandah and offered her a cup of tea.

"It's been so humid," Vicki said. As a light wind stirred the jasmine creeper, which was just beginning to twist around the pole supporting the side of the verandah, we leaned back in our chairs and welcomed the coolness.

"I assume it's been some time since you lived in the village," I commented to Vicki, who was slowly sipping the cup of tea she held with both hands. "Do you return there sometimes? Are your parents still living?"

"Not my father," she said, twisting her long, satin-smooth fingers. "He was with his animals in the cattle camp when he died . . . oddly enough, it was only his shoes made with the hide of giraffe that were left in the dirt . . . and nearby we found his ivory ring. He had been living in a corral. My mother tried to bring him home but he refused. So when the army raid came, my father, because he was blind, could not run anywhere and he was shot. All the cattle were taken. My brother was also shot in the thigh. You know, for us, life is like that," she said, with a suggestion of a smile. "The Jie attack the Acholi, steal their cattle . . ."

"Then of course they revenge," I interrupted, "and the raids continue with the Jie, Lango, Sudanese and the Acholi. The Bakora then raid the Teso . . . and the attacks go on and on. Violence always begets violence!" I said, with feeling. The looted modern weapons from the armory in Moroto made this tit-for-tat raiding so lopsided that the Jie dominated and killed many of the other Karamojong clans and neighboring tribes and took all their cattle.

Seeing I was disturbed, Vicki, who was married to a European relief worker, changed the subject and told me of a Karamojong custom. "Our cattle are so important to us. It bothers the *wazungu*, I know, that we collect their blood after shooting an arrow into the ox's neck, and we then mix the blood with milk and drink. What is left thickens in a bowl so we can roast it. Do you know of our Karamojong headdresses?" she inquired. Before I could say we had collected a few, she went on to tell me that her father had created them as an occupation. "I used to sit and watch him comb and twist, and braid the sisal before he packed it all together with wire," she explained. "Sometimes he painted the wig with red clay and decorated the headdress with ostrich or cock feathers for the men to wear on special days."

Vicki had been given a Christian name when she attended a mission school not far from the settlement she was brought up in. The last of eight sisters and two brothers, she was the only one to have completed standard two.

"My people thought that if a girl went to school she could no longer believe as a Karamojong. They were worried I would forget the traditional way of living," she added. "Like when they kill a goat and smear the blood and the guts on the body when a person is sick." When I didn't react, she told me the witch doctor must be present and the skin of the goat was then put around the person's arm.

"The school I attended was about three miles away," she said. "My mother and elder sister tried very hard to take me away from the school and I had to hide. But they kept on coming, because they did not want me following white man's customs." And although the priest tried to persuade them otherwise, Vicki's family threatened him so much that he finally discontinued paying for her school fees.

"In our custom," Vicki reminded me, "women's work is to do most of the things; to feed and shelter the family. The man just sits, and eats and watches his cows. Now the cows are reducing in numbers and our main food has become sorghum and the beans that foreigners send. These foods do not settle well with my people and they complain of stomach pain."

Rolling her expressive eyes, Vicki told me that, in the days that followed, she resisted the admiration of many men. "For if I slept with a man, my people would no longer allow me near," she said. "I did not marry early, despite my parent's persuasion, but found other means to continue my education in Teso."

"Before a Karamojong takes a wife," Doris Clark noted in the *Uganda Journal,* "he must approach his parents and his brothers," to decide how many cows can be spared for a dowry. (If the total given is less than promised, the girl could be taken back to her own people.) After consuming copious amounts of beer made from sorghum and milk, the men pass around a pipe filled with tobacco to ratify the agreement. The girl is then rubbed with black paint and pushed away to accompany her husband and bear children. Two women bark like zebras and lead the bride to her husband's village, where the door to her house is shut and the fire kindled. The smoke that envelops her is a symbol of her ownership and a welcome to her husband's family. The husband knows the girl is really his wife and can never be taken from him when he receives his cattle. Donning a leopard-skin cloak, knee bells, headdress and a zebra-tail that he wears just above his left elbow, he circles the village, tossing his head and swishing his tail. Finally, everyone gathers into the cattle kraal and the ceremony is concluded with a dance.

Treasures

Vicki had stirred my interest in the Nilotic Karamojong and on a later day, Cole and I took our children on a safari to the northeastern corner of the country she had told me about. Fields of tall grass waved in the wind near clusters of cattle camps in an area with very few trading centers compared to the rest of Uganda. The Karamojong men who strode along the roads carried spears, and sometimes guns, and wore little or nothing at all but an occasional blanket slung over a shoulder. A leather strap covered the sharp cutting edge of the circular knives they wore on their wrists to use as lethal weapons if necessary.

Farther down the dirt road, a group of women wearing skirts made from the hides of animals and bead necklaces and earrings were filling their calabashes from a water pump. We stopped to inquire whether we could fill our jerry cans with water. Timorous yet inquisitive, the women, their hair braided into tight coils and wearing copper bracelets on their wrists, smiled shyly, exposing wide gaps in their bottom teeth. They were particularly interested in our children and waved when we left them to drive up the rocky hill between Kaabong and Kidepo toward a weathered sign that read Fort Wilson.

Prickly cactus and various types of aloe kept us on the path that led to the entrance of John Wilson's residence. John had heard us come up the drive and welcomed us into his cool, mud-plastered abode where, after my eyes adjusted to the dim light, I examined and admired the pottery, spears, lip plugs and traps, along with the other local artifacts he had collected over the years. An accomplished, Scottish-born botanist who was fluent in three languages, including Karamojong, John had spent many years in the Karamoja and was fascinated with the parallels between some of the objects Egyptians used and those still utilized by the Nilotic Karamojong. He had also found remarkable similarities between Hebrew and Karamojong words, such as "echirit," leg bells, in Karamojong, similar to "shirit" in Hebrew; and more than eight hundred Karamojong words similar to Gaelic. For instance, "cro" (to pen cattle) in Gaelic and "aki-ro" in Karamojong; "cur" (sowing) in Gaelic and "aki-cur" (to prepare ground for sowing) in Karamojong.

Convinced certain African cultural elements were the same as those of Asia even before the emergence of Babylonian civilization, John Wilson later set up the Treasures of Africa private museum in Kitale, Kenya, with a rich selection of cultural artifacts he collected during more than thirty years researching Uganda's culture and linguistics. Confident his findings would herald "an academic shake-up," his goal was to help educate people on the importance of African cultural heritage.

John first arrived in the Karamoja with the colonial government in 1949 as an agricultural officer. Referred to as the "Red Baboon" by the Karamojong, rumors suggested that, when Idi Amin came to power in the early 1970s, John had been one of the half-dozen white men who carried Amin on their shoulders. Whether this was true or not, John Wilson was an exceptional character who knew the ways of the Karamojong and was thus able to help them transition from cattle-keeping to hoe agriculture. Employed by Oxfam as an agriculturalist to resettle the Dodoth, who had lost their cattle in raids, he encouraged them to grow more than tobacco and maize on their stony plots.

"The task was a difficult one," he admitted. "Water is scarce and plants wither quickly. Still, some of them have settled to cultivate their land, but more often, the Karamojong earn local currency by chopping down trees to make charcoal . . . which, of course, is no good for the environment. Most of the time I travel in a pickup in the company of armed warriors," John told us.

The Karamojong were an embarrassment to Idi Amin's government. Shortly after coming to power, Amin disregarded the cultural implications and ordered the Karamojong to put on clothes and change their "outdated" customs. But the Karamojong resisted and remained armed and aggressively hostile toward Amin's notorious soldiers.

"Did they fight?" Sanjeeb asked, keen to hear more.

"Of course," John replied, "and they still do."

It was twilight by the time we left John's abode on the rock-strewn hill. And as we drove on to find a place to pitch our tents, the children were full of questions about the man they had met, whom they had found "interesting to talk to."

Kidepo National Park

Grazing land for the Karamojong's cattle diminished in the north of the country, where it bordered the Sudan, when in 1962 Kidepo was turned into a National Park. Although Quentin Grogan, the last of the old elephant hunters, referred to tourism as "the dreadful commercial disease," game viewing and tourism contributed enormously toward

Uganda's foreign exchange and, as a result, conserved the animals. But as Uganda's economy and security deteriorated and government troops and poachers roamed the plains in battered vehicles with automatic rifles to kill for meat and ivory, there was the real danger that much of wildlife would be eradicated.

Aerial surveys showed that almost every species of animal except the Uganda kob were rapidly disappearing. According to *Swara* magazine, between 1973 and 1976, the elephant population dropped from 14,300 to 2000 in Murchison Falls Park and from 2700 to 700 in Queen Elizabeth Park. By 1980, it became necessary to take immediate, urgent steps to focus attention on Uganda's conservation problems in order to save the animals. In 1981-82, UNDP provided funds to purchase vehicles, fuel and guns for the demoralized park guards. Scottish-born Iain Douglas-Hamilton, who was employed to help stem the poaching, was known to dive his Cessna 185 at poachers to disperse and frighten them.

A Pride of Lions

A gust of wind stirred the tall grass as we approached the Kidepo Park headquarters in December 1981. Cole stopped the car near a man in a faded uniform seated at the entrance of the park known to be one of the most beautiful and remote in Uganda, but very rarely visited during the early 1980s. The man, who wore a faded uniform, was surprised to see us and said so, before Cole disappeared with him into a little shed to pay our fee of a few hundred shillings to camp for a few days.

"You must take an *askari* with you with a gun," the ranger advised, informing us he and his colleagues had recently been attacked by Sudanese poachers who crossed the border. "They used heavy weapons and shot at Mr. Douglas-Hamilton's plane," he said, adding his sparse frame to the back seat of the Land Rover, which was already occupied by our three children. The firearms the Sudanese used in their ongoing raiding of neighboring tribes were also used to kill elephants. "Poachers are supplying ivory outside the country," the guard said as we bumped over the gravel track. This illegal trade was having a devastating effect on big game.

It was evening by the time we spread our tent. Dry leaves and twigs crunched under our feet as we pegged the corners of the canvas into the hard soil. At twilight, when the sharp outlines of the distant hills softened into shades of the impending night, warm wind stirred the plains and brought the sweet scent of crushed grass. The children threw sticks on the fire and watched the flames leap and broaden the circle of light as hyenas moved stealthily near. Soon there were enough coals to heat the heavy pan, on which I fried onions, garlic and chilies mixed with beans. The children were hungry and helped me pat out the chapattis, which they devoured as quickly as they came off the pan. When the fire died down, the children were convinced they could practically touch the myriads of brilliant stars that hung close in the celestial calm.

My stomach had been bothering me with what was probably giardia. During the night, I unzipped the flap of the tent to relieve myself, when suddenly a low, menacing growl, inches away, stopped my heart. The unmistakable odor of wild beasts and another intimidating growl roused my husband from his sleeping bag and we peered from the window to watch lions prowl. Frozen with excitement and fear, I heard Cole hiss, "Why are they so close?"

That broke what was left of my composure and I yelled, "*Aaaaaskarriii*!"

Cole joined me with shouts for the watchman as the tail of one of the cats slapped the side of the tent. After what seemed an interminable interval, we heard the rusty door of the small shed twenty meters away creak open and our guard, wearing a heavy military coat and dragging his blanket, appeared. Seeing lions in the moonlight, he bolted back to his tin shed. Cole then jumped from the tent to the Land Rover and drove slowly around the pride of six lions until they sauntered off.

"They were just curious," we told the bewildered children, who rubbed their eyes and asked what all the fuss was about. The guard peered from his shed. After the lions had gone, he asked why we chose to sleep among dangerous animals, then gathered his blanket about him and flopped against the side of the car near the freshly stoked fire. Propping his gun against his knees, he welcomed a cup of hot tea and joined us as we watched for the first eastern light to touch the silhouette of the

Natera Hills.

Grevy zebra, hartebeest, ostrich, gazelle, giraffe, guinea fowl and elephant paraded past our camp that morning. Later that day, a splendid lion with a thick black mane tossed his magnificent head near his tawny mate as Cole drove the Land Rover some distance from where the cats panted in the shade of a low tree. Farther down the road, hyrax scuttled on the great gray boulders outlined against the clear blue sky near the rock escarpment and Grand Katarum lodge, which had not been completed since construction began in the 1970s.

Whirlwinds spun the next morning as we drove toward the Kananarok hot springs near the unmarked border of the Sudan. Suddenly, the guard gripped Cole's arm and motioned for him to stop.

"What, Pop?" Sanjeeb inquired anxiously.

Pointing ahead, toward a thicket of acacia trees, Cole called our attention to a thin wisp of smoke that curled and blended into the pale sky.

"Look, they run quick," our excited *askari* exclaimed, as we cautiously drove toward the dying embers of a recent fire. "They leave their clothes behind."

"Who's 'they'?" the children asked, looking about with wide eyes. There was a sinister feeling about the place. The men who had departed in haste left remains of what we had seen and earlier grieved about—the rotting carcass of a great elephant brutally slaughtered for its tusks. Aware that the poachers were armed, we drove quickly back to our camp.

That night, Cole and the children laughed when I told them I intended to sleep in the Land Rover. But just after I closed my eyes, roar after deep roar shattered the night and three blanketed forms leapt into the car with me. Turning on the headlights, the excited children and I watched through the window as Cole, with disheveled hair and a wild look in his eyes, stood only yards from the curious cats. He threw wood on the dying fire. The lions stared, their eyes glinting glassy yellow in the light, then slipped into the darkness, leaving behind crushed grass and feces.

Many excursions later, the headlights of a vehicle illuminated our faces as we sat around the campfire. "Hello, hello. My name is Paul

Ssali. We rarely see tourists camping in Queen Elizabeth Park these days," the game warden said as he leaned from the window of his Land Rover. Delighted to see a Ugandan conservationist, we suggested he have a glass of beer with us. "I've always been fascinated by animals," he said, before he told us stories of the natural world and of tracing poachers in the aircraft he flew. "It's been a real struggle, but the animals are slowly reentering the park," the good-looking man said enthusiastically. "The elephant population is now over two hundred!" Ten years earlier, there had been more than three thousand elephants in the park, as well as rhinoceros and numerous herds of plains game.

"Oh, by the way," Paul said as he rose from his chair and bid us goodnight. "Just the other day, I saw a female lion with her cubs not far from where you have pegged your tents. So take care."

Encouraged by Paul Ssali's dedication to the cause of wildlife, we watched the headlights of his Land Rover fade into the distance as he drove back to Mweya Lodge, which was slowly being renovated, having been looted during the Amin years.

"Remember Paul said there may be lions near," I called to the children, as they headed for the bush to urinate. As I spoke the words, throaty growls sounded nearby. Petrified, and still pulling up their pajama bottoms, the girls scrambled toward the car while I banged the lids of cooking pots. One of the tents was close to where the growls were coming from and David (my brother, who was with us at the time) and Cole dragged Sanjeeb, who was already in the tent, closer to the car . . . something that caused a great deal of laughter. That night the lioness brought a kill to feed her cubs. The hungry lions chewed noisily on the carcass until morning, when many hippopotami, close to whose trail we had unknowingly camped, returned to the water of the Kazinga Channel. Later in the day, a hippopotamus humped its back in the water, nearly overturning the boat we had hired to motor along the waterway.

Sometime later, we were unhappy to hear that Paul Sseli had been dismissed from service. Paid minimal wages, he was caught carrying fish in his park aircraft to sell in Kampala to subsidize the meager income he received.

Following the Day's Route

It was time to leave our camp in Kidepo.

The children darted about like the grasshoppers they tried to catch not far from the portable canvas "tub" I was bathing in. Their happiness gave me a rush of satisfaction and I made wishful plans to release our little duiker and bushbuck in the Queen Elizabeth Park—sadly, something that never materialized, as Paul Ssali had told us our domesticated gazelle simply would not survive, what with poachers and predators. I had so wanted her to have a taste of freedom. Buffalo dung attracted buzzing flies, and bees swarmed restlessly near the bar of soap I had placed on dry twigs. Leaning back to catch the warmth of the sun, I heard Cole say, "Let's get ready to go," as he began to dismantle the tent.

Pulling on a T-shirt and jeans, I sloshed out the bath water so the canvas could dry in the sun before we packed up our tent and supplies.

The old pre-independence map of Uganda we had showed a track (constructed during World War II by the King's African Rifles) winding down the escarpment northeast of Kaabong. Although the descent looked possible on the map, it soon became evident the road was badly eroded and overgrown with brush. There was simply no evidence of anyone having used it for decades.

"Cole," I said with alarm as we started down the track, "please let's go back."

"How?" he said, as we continued to lurch down the escarpment. He was not overly concerned, and perhaps was even in his element.

He was right: there was no way we could turn around. Strung out and jumpy, I silently scolded Cole for bringing us into yet another impossible situation as we inched around the large rocks of the steep incline to the Lokichoggio plains of northern Kenya. It was a relief to reach the bottom of the ravine.

"But now where do we go?" I asked Cole, who anxiously scanned the undergrowth that stretched toward distant thorn thickets. Overhead, kites dipped and soared, their dark shadows racing on the dry earth.

"Papa, where is the road?" Jessica inquired sleepily. Despite the slow lurch down the ravine, she and the other two children had just woken up from a nap.

"Somewhere, I hope," Cole answered, draining the last mouthful of sweet tea I poured for him from the thermos. "Surely the riverbed we're following will lead us to a track," he muttered as he drove the Land Rover farther and farther east through the soft sand. At one point, we thought of returning to the escarpment we had descended. But instead we drove on, as dik-diks bounced like hares and now and then guinea hens scattered in the dry bush. Tired beyond caring, the children slept with sunlight playing on their features until, at last, a track led us to the border post between northern Kenya and the Sudan.

The army men we had seen ahead were startled at our unusual approach and bounded into shallow trenches with their guns. Not knowing what to expect, Cole stopped the Land Rover some distance away and I held my breath as he raised his arms and walked slowly toward the armed men. "But vehicles do not appear from the Uganda side," the guards insisted, though the evidence was right before them. After some time, Cole talked the men into allowing us to drive to Lodwar and on to our destination—the shores of Lake Turkana. The heat was oppressive and the children complained as we drove through the desolate stretch of country, which was peopled by fiercely independent cattle-herding nomads known as the Turkana, who the Karamojong raided but also considered close to their clans.

It was mid-afternoon by the time we reached Eliye Springs. On everyone's mind was the need to find shade to pitch our tents. Finally, we found a place under a grove of trees near nomadic huts. While the children chased the goats away from our camp, I put the kettle on the gas stove to boil, as we had run out of drinking water. Claiming they would "drink anything," the children sipped the scalding murky water to quench their thirst before they wandered down to the lakeshore to watch naked children swim. Then, when a welcome breeze cooled the air, and before the sun set, we found a goat path we followed to where Turkana women sold their wares under a flowering acacia tree. The women's smiles softened their weathered features as they urged us to purchase the bright strings of beads and bracelets they had made. Bantering and haggling with us over prices, one old woman settled for the amount I offered her and hung beads caked with mud and smelling of rancid butter around my neck. That night, we lingered on the warm sand under a low moon and discussed the following day's route, which would take us south to Kitale on the Kenya side.

The next day we drove on to Mt. Elgon (another of Kenya's extinct volcanoes, which erupted millions of years ago), where we passed women walking alongside the road with kindling strapped to their backs through the rainforest, toward what we had been told was the national park's campsite. Noisy Ross's turacos scolded us as we set up our tents on level ground near a grassy slope that led down to a little stream. Above us, colobus monkeys jumped from tree to tree, their long, bushy black-and-white tails dangling from branches as they paused to watch us with bright eyes.

Just after dawn the next morning, I waved to Cole, who broke a walking stick from a dead cedar tree and beckoned to our two older children to accompany him on a walk to the salt caves. Keeping a watchful eye out for the old buffalo that grazed near the camp, I hung the last wet shirt on a bush to dry. Then Jhorna and I drove to Kitum Caves, where we had agreed to meet the rest of the family. Parking the car in deep shade under tangled vines, we startled a bushbuck, which bounded gracefully past the trickle of water that coursed down the rocks into a pool at the entrance of the caves. The rank odor of bats hanging from the ceiling of the cave permeated the air as we eased our way carefully through the cavern as far as natural light allowed. It was fascinating to see the engravings the elephants had carved when tusking for minerals and salt on the sides of slippery rocks, and the children begged to stay longer. But evening darkened the caves and we hurried them back to the car in the fading light.

The drive from Ishasha at the southern end of Queen Elizabeth Park to Kisoro, the southwestern-most location in Uganda, was in an area known to be insecure. Advised to take care, we drove through the Impenetrable Forest and at the end of the day inquired of the custodian at an abandoned guesthouse where we could pitch our tents. Directing us to a spot under a large tree, he assured us it was a secure place. It was cold and the wind was blowing as we set up our tents at the end of that long day. No one was around, though we discovered that the fireplace had been used in the guesthouse.

"An army officer stayed here for some time," the old caretaker told us as he took us around the neglected premises. "My problem is that I have not received my salary for many months. So I plan to sell these to a trader in Kasese," he said, showing me the books he had stored in cardboard boxes. The books were beginning to mold and I asked the custodian whether I could have two that were authored and signed by George Schaller, the world-renowned conservationist, who had stayed at the guesthouse. Schaller is often called the greatest naturalist of the twentieth century. Through his ground-breaking research on the Virunga mountain gorillas, he influenced Dian Fossey to begin her own gorilla investigations. Schaller's passion for wildlife had established over twenty parks and preserves worldwide.

Yards of colorful cloth and pottery stacked near the side of the road attracted my attention the next day as we drove along the road to Kisoro. There was a market going on and we stopped to purchase some of the crafts for sale. While I selected a few intricately woven baskets, Cole and the children met some men and women of the Batwa pygmy tribe (whose territory spans the mountainous border terrain between the Congo and Uganda), who insisted on dancing for us for a few shillings.

Hoping to see the Virunga mountain gorillas, we left town and took a road toward the forest reserve, through a densely populated area, right up to the steep slopes of the volcano. At the end of the track, the locals directed us to large fig tree they said would be a secure place to spend the night. But as we started to unpack, I kept my eye on a group of unkempt men, who eventually approached us. "You did not ask my permission to stay here," one of the men said. "I am the Uganda People's Congress (Obote's Party) Chairman," he boasted, flailing his cane. "Pay me money."

In an effort to placate the angry man, who reeked of *waragi*, local alcohol, Cole offered him a chair, which he took before renewing his threats.

Unnerved, we quickly packed our belongings into the Land Rover.

Most of them ended up on the back seat on top of the children, who laughed in spite of our hasty retreat down the mist-covered mountain road. Farther along the track back to Kisoro, we followed a sign that led to Traveler's Rest, the hotel once owned and operated by Walter Baumgartel, who was said to have been an eccentric Austrian.

"This place has seen better days," the old caretaker admitted as he walked us around the derelict premises that had once been maintained with manicured green lawns and trimmed bougainvillea hedges. But shadows were lengthening and it was overcast, so we paid for a room that had a toilet attached and put up our roof tent in which to sleep. Torrents of rain clattered on the dining room's tin roof that evening, and we were happy to be indoors as we ate *matoke*, steamed bananas, at a table spread with a cloth that had TR embroidered in dark blue on the corners.

On occasion, we traveled to Kenya and Tanzania to escape Uganda's insecurity and nightly shooting. Camping was superb in the Tsavo bush, where it was not uncommon for an elephant family to spend the night near us. We also camped at the base of snow-capped mountains, and along muddy rivers, marshes and streams. One such spot was under a huge old acacia tree in Amboseli with of view of Kilimanjaro on the Kenya side, where Cole stopped the Land Rover not far from a baby elephant and its mother tearing the bark off a tree.

"This mama seems so very unconcerned," he said quietly.

I wanted a picture, and stupidly walked with Jhorna some distance from the elephant and her infant (others have been killed approaching wild elephants on foot). As I focused my camera, I heard Jhorna say, "Mom, I have a thorn in my foot." Just then, the elephant churned dust, and Jhorna and I sped toward the Land Rover. The elephant stopped meters from the car we had jumped into. Trumpeting a high-pitched warning that sent chills down my spine, she spun around to rejoin her calf.

Jhorna was so traumatized that when I dug out the thorn, which was now deeply embedded in her heel, she hardly complained. The next day, she chose to stay in the back seat of the car while the rest of us walked to the entrance of Tanzania's Manyara Park. Suddenly, with a shout that caused us to turn around, she jumped from the Land Rover, followed by a very large baboon, who was devouring the banana she had been eating. Despite the incident, which could have caused Jhorna more than unsettling anxiety, we laughed heartily as the baboon loped into the forest with his loot.

The next day, Cole drove the Land Rover up the slippery muddy track to the Ngorongoro public campsite on the edge of the largest unbroken caldera in the world. Perhaps because tourism was down and the border between Kenya and Tanzania had been closed following the collapse of the East African Community in 1977, we were surprised to find we were the only ones camping in that spectacular setting. Since it was the night before Christmas, we decided to treat ourselves to a proper dinner at the lodge.

"Enjoy the view," the old waiter told us as we pulled our chairs to the side of the verandah that overlooked the crater. Although it took time for the food to arrive, we looked forward to what the staff assured us would be a feast worthy of the occasion. But when our dinner finally arrived, the children did their best to hide their disappointment at the sight of the tough little browned game fowl and boiled potatoes.

The hours of darkness brought strong winds that blew incessantly at our camp that night as we watched a solitary bull elephant quietly walk by. Despite the blustery weather, we were exhilarated with the wonder of it all and huddled together near candles that wouldn't stay lit, to celebrate the season with hot chocolate and banana cake. The next morning, we drove past fields of yellow wild flowers that sloped down to Maasai *manyattas*. A series of tribes had herded their cattle in the area before the Maasai (who were allowed to drive their cattle into the crater for water) settled there only about two hundred years ago.

Thompson's gazelles bounded near us as we lurched over the rutted road to Oldupi Gorge, the site where Louis Leakey's wife, Mary, searched for fossils and a wealth of hominid and animal remains. Fascinated with the pictures of skulls and bones of Zinjanthropus displayed in the small museum at the site, which were known to be 1.75 million years old, the children asked many questions as we sat on the edge of the gorge and ate our lunch of leftover chapattis and tuna fish. Although the children were reluctant to leave Oldupi, our intention was to reach Lake Ndutu that night and pitch our tents near the late Dutch photographer Hugo van Lawick's base camp. We drove on through the Serengeti, white volcanic dust blowing over the parched ground and swirling around the Land Rover, as Cole kept us entertained with the story of the remains of Lucy found in one of the lowest points on the earth's surface, which dips to more than 328 feet below sea level.

An Unfortunate Incident

Water and diesel slopped in the jerry cans strapped to the front and roof of the vehicle on another day as we approached Moroto. It was December 1984, and we were looking forward to Christmas in Kidepo National Park in the far northeastern corner of Uganda. Conscious that we were in an area of chronic insecurity, Cole drove to the World Food Program (WFP) office and switched on the mobile radio to find out more about the situation. As hot, dry wind blew through the open windows that afternoon, we were barely able to hear Mons Swartling say through the static that he had been shot in the arm by Karamojong raiders and was driving north toward Kotido as the road to Moroto was blocked by armed raiders. Concerned about his condition, Cole advised him to put a tourniquet on his arm, and warned him to loosen the cloth from time to time. While Mons kept us updated, Cole relayed messages to the UNDP security office in Kampala, until Mons managed to reach the small hospital in Kaabong, north of Kotido. There the Italian Dr. Forney did what he could, before Mons was evacuated by army helicopter to the Nsambya Catholic Hospital in Kampala where WFP and UNDP staff took over.

Unsettled by the incident, we decided to change the order of our day, and drove south around the slopes of Mount Moroto to a remote spot, where we set up our tents near a herd of impala. Rising early the next morning, we took the bush road that led to the border linking southern Karamoja to northern Kenya, through the land of the Pokot, north of Kitale. That night, we celebrated the birth of Christ with songs and praise around a fire on the cool slopes of Mt. Elgon, which bordered Uganda and Kenya.

The next day, we reached Kampala. Although it was New Year's Eve, Cole and I left the children on Mbuya Hill with Pelesia and drove to Nsambya Hospital to see how Mons was doing. The smell of rotting flesh nauseated us as we opened the door to his private room. Mons was soaked in sweat and only partially coherent, due to the drugs and sedatives he had been given for his pain. Alarmed, Cole phoned the UN doctor, Ray Gibbons, who came straight to the hospital, along with the Russian orthopedic surgeon treating Mons. That evening, the doctors amputated Mons' arm. Soon after, he was flown on an emergency flight to Sweden, where doctors removed another few centimeters of his arm to save his life.

Chapter XIV
Lest History Repeat Itself

ONE FRIDAY AFTERNOON, Malayamu, the wife of a Ugandan friend who worked at the university, and I were seated on our verandah on Mbuya Hill, enjoying the light breeze that swayed the branches of the flamboyant tree. I watched the sunlight play on the scarlet and pink bougainvillea spread over the side of the tiled pavilion, and marveled at the tiny sunbirds flashing among the flowers. Nearby, pygmy kingfishers and plantain eaters perched noisily among the hibiscus blossoms. In the distance, the lake glistened behind green swamps nestled between gently rolling hills.

Malayamu said, "We used to listen to Radio Katwe (the rumor station that spread the local news) to find out what stage the Tanzanian troops were in for taking over from Idi Amin." Then she leaned back in her chair and laughed heartily, before telling me she had rejoiced when the troops came to liberate Kampala in 1979. Her laugh was infectious, and we both continued to express amusement at what was impossible to comprehend and difficult to explain.

"Those Tanzanian soldiers wanted watches and TVs, but me I went for sugar. My husband and I carried eleven bags of sugar from the brewery that day. We kept two and gave the rest to our people," she admitted.

"How did you carry it all?" I asked.

"With difficulty," she chuckled, and added, "Then one man fell, under this fifty-kilo bag. We looted from him as well," she said, still amused.

As she talked, my gaze took in the high-rise apartments in the valley below, where, only a few nights earlier, people had been subjected to armed robbery and probably much more . . . all the while thinking Malayamu's words followed actions no longer under control.

"Can you imagine?" she continued, twisting a lock of her hair. "The whole first week of liberation, the looting went on. Soldiers shot the locks off the doors and not only those on abandoned houses; occupied ones were looted as well, along with shops and offices. My poor friend did not even have as much as a spoon or a needle left." Finally remorseful of what she had revealed, Malayamu became quiet at the recollection of the plundering. I thought with sorrow of the background of events, which had culminated in conditions that had not changed. Those we knew were preoccupied not only with their safety, but also with how to subsidize their meager incomes in order to purchase staples such as cooking oil or even a loaf of bread, which were often overpriced . . . if they were even available.

We Did What It Took to Survive

Other Ugandan friends lowered their heads and spoke in hushed tones when reminiscing of the horrendous crimes Idi Amin and his henchmen had committed. Amin had told his soldiers that "the gun is your father and mother."

Imagine the stories the marabou storks could tell us, I told the children one day, as Sanjeeb observed the large scavenger birds preening their feathers on top of the street lights near the Speke Hotel. The storks, with their over-sized bills and drooping pink throat-pouches, had surely witnessed gruesome events. Farther up the hill stood the derelict buildings of Amin's former State Research Bureau (SRB), where people had been tortured and killed behind the tall hedges.

"Amin's SRB was notorious," an anonymous student at Kampala's Makerere University narrated in a circulated report. The student's friend had been detained at the SRB for over a year, but fortunately came out alive, together with several others, when Kampala fell in 1979.

"I got curious," he begins. He then joined a group of others to see for himself the SRB, which was a place synonymous with the torture and terror described in foreign newspapers but not talked of much in Uganda for fear of ending up there oneself. "Outside the building," he continues in the report, "was a courtyard which appeared to be a parking area . . . A Tanzanian army officer guarding the premises said the poles were used to tie prisoners to be flogged or branded with hot iron in order to obtain information. To the left of the main entrance, was a passage leading to what looked like a urinal. Shiny pink liquid covered part of the floor and a few ladies' bras and underpants were left here and there. This was the place, we were told, where women used to be killed and what we were looking at was clotted human blood. The place was putrid and smelled like fermenting cow dung and rotten meat. I held my throat."

Another room, he reports, had photo albums of people from different countries of origin and yet another room had a large collection of *Time* and *Newsweek* magazines that he helped himself to "as there appeared to be no restrictions by the Tanzanian official guarding the room." He found piles of personal mail, journals and newspapers in the office, and was confidentially informed that most of the important documents "including top secret Tanzanian invasion plans, had been sneaked out by some British reporters who visited the place soon after the so-called liberators had chased away or dispensed with the Amin men.

"Several other offices we visited had a shirt or a pair of underpants or a belt or a pair of trousers lying about and there were stains of blood in these rooms, either on the floor or the walls." In Uganda, the writer states, "things have a tendency to assume a mysterious shroud. No bold historians wish to put pen to paper lest history repeats itself too soon and they become victims of a regime that has regained power."

The last room they visited was the infamous dungeon or cellar "where our student friend had been kept for weeks on end without any light. He looked a mere bag of bones, a skeleton, when his liberators released him. Some Amin soldiers had tried to kill all of the people in the underground cellar just prior to the arrival of the liberating army by throwing grenades. A few people died but others survived the blasts. We attempted to go down this stench filled cellar with the help of a hand torch but failed even to see the steps we were expected to descend on. We gave up this venture and returned to Makerere rather disappointed at this failure of our expedition. I would have gone back to visit the place again had my wife agreed to come with me. But she was too horrified to accept the suggestion, so it never happened," he concluded.

Following the overthrow of Amin, an elderly academic by the name of Yusuf Lule headed the compromise coalition formed to rule Uganda in 1979. But Lule remained in power for only sixty-nine days. In an attempt to control the chaotic political situation in Uganda, Binaisa, the British-trained lawyer who replaced Lule after the latter's unstable government collapsed, planned for a general election in June of 1981. By May 1980, factional conflicts came to a head and the military commission headed by Paulo Muwanga seized power, deposed Binaisa and returned Milton Obote to the presidency and commander-in-chief of Uganda's armed forces through elections.

"It was honestly so bad for our children during the Obote years," Rose told me one afternoon at the Uganda Craft's coffee shop. "Our children heard of their friend's parents getting killed and often asked me whether anyone would come for Daddy. My worry," she continued, "was for my husband, who, as the head of personnel at Makerere University, had access to who was hired and fired." Confirming her worry, one day men visited her husband. They produced identity cards from the Ministry of the Interior and warned him not to dismiss northerners such as the Acholi and Langi from the faculty.

"Then we really lived in fear," Rose recalled, shaking her head. Frightened a knock on the door might mean the rap of death, she drew maps for their children so they could find the homes of relatives: "in case we went missing or were not home when they returned from school. We instructed our children who was an uncle and who was not," she added with a sad smile. "Of course it was an advantage to be a member of Obote's Uganda People's Congress (UPC) party when campaigns for elections started," Rose explained, "but my husband detested politics. Then one day, someone told me they had seen my husband's file in the president's office and in it, a notation saying, 'find a replacement for this man.'" Fearful his life was in danger, Rose sought the advice of a friend, who advised them to begin attending political functions.

"Those UPC T-shirts we then wore turned our lives into something that was not genuine," Rose said, shaking her head. "We became so unhappy, and my children no longer had good appetites. But we did what it took to survive, although we were not being true to ourselves. This is the kind of thing that made us sympathize with Museveni's guerrillas," she ended bitterly. "To get rid of the lot oppressing us."

Convinced the election in December of 1980 had been rigged, Yoweri Museveni took his men to the bush to launch a guerrilla war, and claimed in 1981 that one-third of Uganda was under his control. Obote responded by expanding his dreaded security and intelligence services and revived the feared General Service Unit (GSU), which he concentrated in the Luwero Triangle—the area between northern and western Uganda. He also increased the size of his largely Acholi and Langi Uganda National Liberation Army (UNLA) under Tito Okello, which the British Army trained and equipped at the barracks in Jinja.

It was then that lists of wanted people were passed around and the killings systematically started, another friend, Sarah (not her real name) told me. She had spent most of her childhood in the palace of Toro, where her father's job was to guard the king. But rather than elaborating on those difficult days, she told me of other things. Everyone in the Kingdom of Toro, she recalled, was invited to the ceremonies that were taken very seriously by King Rukidi, who she fondly recollected had favored her. Known as a humble man with an imposing manner, the king was interested in preserving his people's traditions. "Men came from every part of Toro to repair the grass roofs of the huts and the fences around the royal enclosure for the anniversary of the king's coronation," Sarah told me. And after the king presented his annual address to those who were invited, there were all sorts of ceremonies and rituals, followed by dancing that went on for days.

"It was an important tradition for us to remain virgins before marriage," Sarah told me. "So when I reached puberty, it was the age for me to be given a virginity test. I felt so awkward when my father's sister put her finger up my vagina to see whether the fence had been broken." Sarah was then closely watched until her husband-to-be was given time to talk with her parents to negotiate marriage and plan for the celebrations that followed.

The daughter of King Rukidi III, Elizabeth Bagaaya, and her brother, Patrick, came to dine with us, along with other guests, on a later evening when Ugandans were beginning to return to their country. She was as elegant and articulate as I had anticipated, and after dinner, Elizabeth signed for us her book *African Princess*, in which she tells of how she qualified at Cambridge University as East and Central Africa's first woman barrister. She then became a fashion model in Europe and New York, after which Idi Amin appointed her as his roving ambassador and Minister of Foreign Affairs. Dismayed and disgusted at the rumor he circulated saying she had sex with a white man at Orly airport in France (most likely because she had rejected Amin's advances), Elizabeth told how Amin's allegations backfired, and he ended up ridiculing himself by instigating worldwide protests, appeals and warnings from heads of state to women's organizations. After that, she had good reason to fear for her life and escaped over the border to Kenya, and later to the U.K., where she received apologies as well as a large settlement from various international newspapers for the false sex claims that had been printed.

The Kingdom of Toro was Professor Kajubi's primary topic when he addressed a group of international women I had gathered. Explaining the kingdom as part of the territory of Uganda occupied by peoples of diverse ethnic origin and different cultural backgrounds that were referred to by the Europeans as "tribes," he wondered why the term "tribe" was not applied to people of different ethnic origins in Europe, such as the Irish, Welsh or Scots. He concluded that "if a nation is a single people traditionally fixed on a well-defined territory, speaking the same language, and preferably a language of its own, possessing a distinctive culture, and shaped to a common mould by many generations of historical experience . . . then our so-called tribes are full-fledged nations!"

Pandigarri and Stolen Goods

Most mornings on Mbuya Hill, noisy turacos woke us to peaceful dawns, despite the gunfire that frequently disturbed the night. Daytime shooting usually meant someone was being hijacked and robbed, or a *pandigarri*, or round-up operation, was taking place. Although a broken lock certainly wouldn't keep out determined intruders entering our house, the slide lock on the back door needed repair and Cole sent Christopher home to fix it. As soon as he entered the gate, the goat he had given us sauntered up to him. "Does she remember me?" he asked, scratching her head.

"She's fond of attention," I told Christopher as I led him to the kitchen door. "And how have you and your family been?"

"It has not been good for us," Christopher told me as he took apart the lock. "Last week a grenade exploded in our *shamba* and the sound was too, too strong. It finished all the anthills and beans my wife had been feeding us with. After that," he continued, "I was rounded up like a stray dog and put in a truck with other men near my home on a *pandigarri* sweep. I felt like running, but the soldiers had guns."

Accused of helping the guerillas, the men were taken to Makinde Barracks, where they were lined up and searched to see if they had money.

"Even pregnant women had to expose themselves to show they were not hiding arms," Christopher said with disgust. "They put us in an underground room where it was so dark, with no ventilation. No toilet . . . and we were beaten."

"Did anyone know where you were?" I asked.

"My wife found out. After bribing the guards, she brought me milk and *posho*. We had not received food and water for two days, and we shared the porridge. We were beaten every morning and evening. If anyone got food, we all shared it. Men came and went. Some were taken out at night and killed. The women were used by the soldiers and some of them were made wives. Three men died in our cell, and an army man tried to cut me with a *panga*," he told me.

"Do you ever see any of the men who were jailed with you?" I asked.

"Of the sixty or so men I was with, I still see four of them—but one has gone mad. Then one day, there was a robbery of an *mzungu*'s car," Christopher continued. "And more suspects were shoved into our cell."

"How did you finally get out?" I asked, hoping it hadn't been a

UNICEF vehicle.

"My freedom was bought with 500,000 Uganda shillings my sister brought from selling our possessions," Christopher told me. "After my release I had heavy dreams for many weeks," he said, shaking his head as he completed his story, "but then, slowly, slowly, I returned to what I am."

"What you are is what I see," I responded sympathetically.

"But for you, you are safe," Christopher added, as he stuffed his tools into a cloth bag. "People will not come to loot your baskets," he said with utmost sincerity, referring to the local handcrafts displayed around the house.

And somehow I managed to smile.

It was probably a good thing we had no electronic equipment in the house, as much of the theft we heard about was for such things. I prayed to God that we would be spared.

"I wouldn't like to alarm you, but we had a robbery last night," an acquaintance of mine who lived behind us on Mbuya Hill said one day when she dropped by to see me. We had met at some function or other, and I knew she was involved with the political side of things. Accepting my offer to have a cup of tea, she handed me a package wrapped in newspaper, saying, "This is for you."

"These are beautiful!" I exclaimed when I opened the box to find six exquisite wine goblets. Smiling at my pleasure, my friend told me her military companion brought her "beautiful gifts" from time to time. My question then, of course, was where he had obtained the glasses. The thanks I was about to offer caught in my throat, and I received the looted crystal with reluctance.

Another friend, who had resided in Uganda since 1960, wrote about the dilemma she felt concerning the matter of stolen goods and the challenge of being a parent in Uganda in a story in *War, Violence, and Children in Uganda*, which was compiled by Cole and Magne Raundalen. Spirited, and a self-reliant academic, English-born Josephine Harmsworth felt she participated in corruption when she purchased what most likely were stolen goods at a low price. The second wife of a Ugandan police officer, Josephine now lived in a flat in the town of Jinja.

"We belie our expressed beliefs," Josephine wrote. "In what sort of society are we bringing up our children? Of what relevance are the things we are trying to teach?"

It seemed that corruption was a progressive disease that involved all of us.

As Victims

Incidents of civilians subjected to horrendous atrocities were not infrequent in the disturbed area near Kampala known as the Luwero Triangle, where up to 300,000 innocent people were killed.

The following story involves the private agony a young sensitive girl was forced to endure. Her story began in 1983, when clashes between the guerrillas and government forces were at their worst.

"They used to kill with knives," Philamena told me, speaking slowly and painfully. Her expression reflected dark despair. "Others were ordered to pierce themselves with thorns. People, no matter their age, were made to sing and dance while naked. After this, people who were to die were randomly chosen and those for torture were left behind. And we decided to hide ourselves in the bush. But we escaped across the lake and walked into Bulemezi, where we stayed for three months. But this place also got disturbed. So we ran back to where we had come from, staying in the bush. Sometimes we went for days without eating. Whatever we found, we ate without cooking. The situation worsened and the military men began raping girls and old women. My father was gunned down after refusing to obey their orders to perform sexual intercourse with me, his daughter," she revealed, her eyes downcast. "My brother was also killed when he refused to do the same. After all this, my young sister and me were taken to Lubiri barracks near Kampala. My sister was then taken to Fort Portal barracks, and I remained alone and always kept wondering where my mother and other brothers and sisters were."

She stopped and twisted her hands, while my whole being ached for her.

"Me, I was known as 'guerrilla girl,'" she went on. "I really suffered, and came to think that God never existed. I wished I had poison. They used me—a different soldier every night," she said, crying quietly, her words growing softer and softer.

Finally when this girl-child was able to escape, she found members of a church who looked after her, and helped her renew her faith. Full of admiration for her fortitude, I heard her say, "I praise God. I will continue to praise Him more, though I do not know the whereabouts of any of my near family—not even my extended family, I will praise His name."

Faith

The wind moaned through the high arches of the Anglican Church at the top of Nakasero Hill. It was another Sunday and our family of five was seated on one of the worn pews for the evening service. Normally, an elderly man played the keys of the organ with concentrated purpose. When it came to communion, it amused the children to watch him hold the cup longer than usual to his lips. Heavy faded crimson curtains hung on either side of the small stage on which Christine and her young and hopeful group lifted our spirits with the promise and love of Christ and the vitality of song amid so much hurt. Their spiritual senses sharpened, the faith of these Ugandan brothers and sisters was like trees rooted deeply in the soil.

"There is hope of a tree if it be cut down," Christine read from John 14, "that it will sprout again, and that the tender branch thereof will not cease . . ." she continued in a clear, resonant voice, until tears quenched the fire in her eyes. Turning to her choir, she directed them fiercely into heartfelt song, accompanied with stamping feet and arms raised "higher, higher, higher." Christine then invited those of us gathered to stand up and join the choir, to "push fear and the devil down . . . lower, lower, lower . . ." At the close of that emotional evening, we rose with prayers for God's presence to provide firm footing and the assurance to help us see glimpses of truth through all the pain.

It turned out we needed the extra assurance, because the next day

Cole was once again called to sit opposite the angry vice president in the drawing room of a large house next to the church on Nakasero Hill.

"If you continue to give press reports about Luwero that criticize this government, I will deliver your body to the border in the back of a Land Rover," Muwanga shouted at Cole. The vice president, who was also acting minister of defense, shook his fist at Cole to accentuate his words. His rage that day was because Cole had spoken to Bob Gersony, the State Department consultant, concerning civil war in Luwero, who, after evaluating the human rights conditions in Uganda, had submitted a report to the United States Congress. The BBC picked it up from there, but Cole was quoted in the report.

Rising from his seat, Cole stood tall and mustered his courage. He retorted that UNICEF had a signed basic agreement with the government to allow them to be involved with the women and children in every part of Uganda—wherever they were in need. Then he left. But the vice president's warning left Cole in a disturbed state. And that night, seated on the crest of Mbuya Hill, Cole and I talked through the possible grim consequences.

Luwero had once been a rich coffee-producing region and was now the scene of guerrilla warfare against the army that overwhelmed the people and caused death and despair. Jennifer had been part of the turmoil. Young and severely traumatized, Jennifer told me her story as we sat together on her uncle's porch.

"Soldiers started getting around in the villages, disturbing, and trying to arrest men, calling them guerrillas," she told me, holding my gaze with sad dark eyes.

"We ran into a *panya,* rat path, to avoid the soldiers. My mother, father, brothers and sisters and me reached the Singo area, where we built small huts from grass and sticks and looked for food in abandoned *shambas*. We were just trying to stay alive."

Jennifer, who had been born in Luwero District, studied in the village of Kyamwya until primary seven, when the trouble started in July 1981.

"That day, my father ran in one direction and the other children ran in another. We three girls were taken to the camp Makokoto, which was manned by soldiers. At that camp, of the eight people arrested, seven were just killed. The soldiers used to rape women. One girl . . . she was . . ." Here Jennifer broke off and turned her head to look at me. "I think she was pregnant," she whispered. "She was raped so many, many times before she died. Then another man wanted to take her by force, but she struggled and he stamped on her chest and killed her. We stayed six months in that area. The girls and their mother were then moved to Mukomayro, where they remained for a week before they were arrested because "we were termed guerrillas." Pausing, Jennifer rubbed the tears from her eyes before she continued. "On one occasion, those men wanted me. My mother had to defend me, and in the process I was cut on the neck with an ax that left a deep wound," she said, bending her neck for me to see the scar. "Before, the camp was a school, but now it was overgrown, as all the children had left. Soldiers used to take off the iron sheets and tried to destroy the structure. They made us captives cook for them and had us help to destroy buildings for firewood. There were some Baganda soldiers. They didn't want us to communicate with them in Luganda, so we spoke to them in Swahili. They saved us. After three weeks, there were more captives and at the same time they got more soldiers from Mbarara, Arua and Bombo. They also brought many cattle from wherever they looted them. The soldiers who came from the other camps wanted to have us all killed, accusing us of being guerrillas.

"At that time there were about thirty of us. At least they tried to plead for the elders in the group, but the soldiers took twenty people and killed them near the river, dumping their bodies there. Ten of us were left. Among them my mother, sisters and me. After that we were always terrified. We tried always to be humble, cook for them and to obey. The soldiers who were there first disagreed with the new soldiers from Arua, who were then moved to a different area. Soldiers used to go to areas especially where there was coffee stored. One time they brought a man who was a preacher and three children. When they brought them, they were all killed near the camp, and again we had to have great fear. Those people had been trying to escape and were thought of as guerrillas."

Among the soldiers was an Ankole captain who overheard the

others saying they wanted to kill everyone. The captain then moved the girls and their mother to Bukuya, Gombolola headquarters, where they were left in one room with three other prisoners. The Ankole captain gave them food, but when he had duties in Mubende, the other soldiers beat them with the butts of their guns. One man's leg was broken, and when the captain returned, he had the women transferred to Mityana.

"When in Mityana," Jennifer continued, "soldiers came and asked us where we came from and we lied that our husbands had been killed because if we had said we didn't know, they would have killed us." A kind captain then gave them a letter to travel to Kampala because he knew they were innocent.

"We got on a bus and traveled to Kampala. We went to the Kasubi area and an uncle advised us to contact the Red Cross to find our relatives. At the same time, my father put an announcement on the radio calling us and this is how we became reunited. God is great," she said. Her story ended, but it haunted my thoughts for days.

Murchison Falls

Soon after Jennifer told me her story, Cole drove our children and me down the formidable stretch she had told me about. Few traveled the most direct route to Chobe and Murchison Falls, which led through the Luwero region. A short distance from the outskirts of Kampala, men dressed in faded camouflage uniforms patrolled the edge of the road near a large pile of freshly dug earth. Expecting the worst, I turned in my seat to make sure the children were occupied with a game of cards, and looked out the window to see military men supervising men in rags digging a large hole in which to throw the bodies that lay in a tangled heap. Sick with revulsion, I glanced at Cole, who drove quickly past the grotesque sight and on toward what had been rich farmland, but now was overgrown and neglected. However, clusters of banana trees and cassava plants still grew next to winding paths that led to the interior of the bush, where death and destruction lurked.

The boxes containing pots, pans and kitchen utensils for camping purposes rattled in the back of the car next to a carton of powdered milk, cheese and assorted biscuits, bags of rice, pasta, tea and fresh vegetables. Having driven 217 kilometers from Kampala, the last portion of which had been on dirt roads, we slowed down near the Wairongo gate, where men dressed in tattered uniforms recalling more respectable times sat lethargically near the park's entrance.

One of the men, who had been picking his teeth with a long twig, rose to pull aside the pole that blocked the way through the forest ahead. The air was warm and clammy and we kept the windows open until tsetse flies forced us to close them. The annoying flies left large red welts, which stayed irritated for days. It was now afternoon, and the late sun brushed the grassland with gold as we drove down the track until we reached a small clearing. Pleased to have reached our destination, the children bounded out of the car and at once recoiled from a bunch of dried bird feet tied to strings dangling from the branch of a fig tree.

"Juju," Cole said, assuming the objects had something to do with witchcraft.

Evidence of poachers, I thought, avoiding the twisted shapes of fur and bones that swung in the breeze. Nearby, we could hear a steady rush of water that cascaded under the limb of a large fallen tree and we followed the trail that led up the hill. Parting branches to reveal the view, Cole and I caught our breath in awe and kept the children close. Sprays of mist rose from the churning water that plunged from a narrow chasm to thunder into the Nile below, leaving clumps of foam swirling in the river.

But dusk was approaching and we hurried back to the clearing, which was infested with bugs. It was now a question of where to camp. Without a lot of thought, Cole drove the Land Rover straight up the hill to the summit, where we unpacked our tents and supplies. Darkness fell quickly and the children and a nephew who had accompanied us found blankets in which to wrap themselves against the chill of the evening. Nearby, an owl hooted from a tree and we talked of superstitions linked to the dried feet we had seen as the children crept closer to the fire and I strained my eyes toward unidentified sounds in the dark night.

At dawn, Cole, our nephew Eric and Sanjeeb walked with fishing reel and lines to the lower falls. Later that day, they returned with faces burned by the sun and elated with what they'd seen—evidence that convinced the girls and me to join them the next morning. Vultures swooped in circles over half a dozen bloated hippopotami floating in the whirlpools far below us as we walked down the rocky path. Poachers had cut away the hippo teeth and thrown the carcasses into the river. It would take time for the swollen hippos to be punctured so that scavengers could feed on the exposed flesh.

Poachers and soldiers had decimated Uganda's wildlife. Other than hippos, there were very few animals left on the southern banks of the river. Certainly there were no other tourists visiting Murchison Falls, which had been a prime destination among Uganda's parks. The children scrambled ahead of us on the path that wound steeply down the side of the hill to the water's edge. They had already stripped down to their underpants to play in the Nile water and I watched them closely from the rocks in the sun near to where the men fished. Suddenly, shouting for someone to help him with the gaff hook, Cole hauled in a giant Nile perch. Cole packed a portion of the fish to carry back to curry for supper and left the remains for the crocodiles that waited with protruding eyes near the banks of the river.

Chapter XV
Uganda Crafts

Surely There Will Be a Harvest

WHILE DRIVING THROUGH Kampala on a back road toward Nakasero Market one day in 1982, I was pleased to see a small shop that sold crafts. Ever since I can remember, I've admired anything made with natural fiber. The display in our house proved that, as the floors were spread with jute carpets, the furniture was made from cane and water hyacinth and I could never have enough baskets. The woman minding the shop was pleased to see me. Greeting me warmly, she introduced herself as Betty Kinene. "But I was born Betty Nalubowa in Nabutundu in Bugarere," she said with a smile. We struck up a conversation about her products. Leaning on her metal walking stick to balance her crippled leg, Betty described the old woman who made the basket I had chosen. "She travels sixty kilometers to sell what she makes. Otherwise, if the women do not sell anything, they cannot live," she said. The crafts and baskets she sold were traditionally used for some purpose or the other, rarely as decorative objects.

One of twenty-seven children born in 1952 to a Buganda man with three wives, Betty had been infected with polio at the age of three. "Children in the neighborhood used to clap their hands and call me *Omulema*," she said. "I had to cry a lot, until one day I met a woman who was *completely* paralyzed." Pausing reflectively, she added: "That woman could not walk or move. From that day on, I stopped crying."

The next time I went to purchase something from Betty, I found her in another room up the same street under the derelict Imperial Hotel—"because the other shop burned down," she explained, with no comment as to what or who might have caused the fire.

Some weeks later, I was taken aback to find Betty selling crafts in yet another room farther up the street, having been chased from her previous shop. Justifiably upset, she objected to the man who, she said, "thought the system I used was a simple one and passed behind doors to see the landlord. So they had to chase me away. For where there is success in this country of Uganda, for them there will be a harvest," she said tearfully. "Now what can I do? My husband was recently killed in Luwero (the notorious killing fields) and I am now the sole provider for my eight children. The responsibility is mine."

Snapping her caliper into place to support her affected leg, Betty accompanied me to the car I had parked to one side of the potholed street.

"Sincerely, many suppliers depend on me," she said referring to those who brought her what they had made. Pleased that my friend had found her way around a situation that offered so many obstacles, I drove away.

Soon after, a well-dressed man walked through the door of Betty's shop. Opening his briefcase, he handed her a letter that said the shop had been allocated to him and that she should vacate the premises by 12 noon. "The time then was 10 a.m.," Betty told me later. "And I'm sure that man thought for him there would be a big harvest. So I had to close the place," she said, wiping the tears from her eyes with the back of her hand. Betty led me to the door of her former shop and tugged at the large padlock. "See, he put a lock over mine," she sniffed, turning her face to hide more tears. "What now can I do?"

Incensed at the unfairness of it all, I suggested we take the matter to court. Betty managed a weak smile. Wearily shifting her crippled leg, she reminded me: "You forget. I don't have rights. I'm a woman—I have no money."

I knew she was right. But I also knew that something had to be done.

The following day I walked back over the broken sidewalk and found Betty standing forlornly outside what had been her shop. Blowing her nose into her handkerchief when she saw me, she wailed: "Even they will not allow me to remove my things. How shall I pay the suppliers? What is to become of my children?"

Not knowing what else to do, I took Betty home for talk and tea. And when she was ready to go, I packed milk powder, margarine, tea and bread to take to her children before I told her, in words stronger than the confidence I felt, that we would find funding for her to begin again.

The donor's response to the proposals we submitted advised us it would not be feasible to run a profitable nongovernmental shop for the disabled and crafts persons in Kampala. "Who would buy the products?" they said. However, David Campbell, the Nairobi-based Field Director for Oxfam (who had succeeded Cole in Bangladesh and had a long history in India as a tea planter), thought it was a great idea and responded positively to our request for money. Delighted to have Uganda shillings in hand, Betty and I began our search for those who made crafts.

We drove down a bush track slippery from rain south of Kampala, to where we found men grouped in a clearing. The men were engrossed in a game of *omweso*, played with seeds and holes dug into the ground. Inviting us to share the *mwenge bigere* they were drinking from dried gourds, they laughed tipsily and directed us to a woman stirring the brew of fermented bananas in a wooden canoe with a large stick. We further amused the men by purchasing two of the cracked canoes the woman no longer used as containers, which I intended to use for plants.

"Where will we find those who make mats and baskets?" we asked the men, who told us to go in the direction small children had fled like duikers. Following the track through the forest, we eventually found a man with a kindly face and white beard seated next to an enormous dormant anthill he had hollowed out to sleep in and keep his meager supplies. "When it's wet, I close out the rain with a mat," he told us cheerfully in Swahili, as Betty translated. "The baskets I make," he said, showing us the sturdy pile stacked behind the anthill, "I sell, which gives me what I am." Then offering us a mat to sit on, the old *mzee* told us he had journeyed many miles from a far-off place in Zaire. "But it is hard," he admitted, delighted that Betty and I purchased all the baskets he had made. Word soon spread that we had paid cash to the old man, and villagers followed us down the road until we agreed to look at what they had for sale the next time we visited and promised to bring them sugar, soap and blankets.

Rain washed the windshield when, some years later, Sylvan pressed the accelerator with his bamboo cane (his right leg was rendered useless by polio) and drove us to where the *mzee* we had met near the anthill now lived.

"He talks of dying," those residing near him reported.

Sylvan stopped the car in a clearing, and Betty and I stooped low to enter the door of the small mud hut. The old man had built the hut with the profits he earned from selling baskets to what we had established as Uganda Crafts. It took time to focus our eyes in the dim light, but when we did, we found the *mzee* lying on a thin blanket on the dirt floor, only half the person he had been. Slowly turning his head to see who had entered the room, he recognized us with pleasure and with a raspy voice repeated, "*Malaika yange*," my angels, over and over. Wiping away our tears, Betty and I requested Sylvan and the woman who had followed us into the house to carry *mzee* into the warmth of the sun. Leaning the frail old man against the side of his humble dwelling, we coaxed him to drink the tea with sugar and eat the bread we had brought.

"I am encouraged," the gentle old man told us. "Now that I have seen you I will again become strong."

Confident his health would improve with the supplies we gave him, we left the old man, who lifted his arm weakly to wave in response to our goodbyes. Two weeks later, when *mzee* was strong enough and ready to leave his little house, we shifted him to the Uganda Crafts village near Lake Victoria, where he had access to clean water, enough food and health care.

Craft Shop

Late in November 1983, the contractor nailed the last strip of banana fiber firmly into place. The renovation of the spacious room below UNICEF was complete. Bright with natural light, the former Uganda Teachers Association library lent itself to a splendid display of arts and crafts Betty and I had collected and stored in the bedroom of our house until Uganda Craft's opening day. Beaming with undisguised pride, Betty said our efforts had satisfied her expectations. Before we hired others to work for us, Betty handled the accounts and I selected and displayed what was sold. Very soon, customers came and sales increased . . . and later soared. Our policy with quality control and payment to the women on receipt for what they made enabled them to pay their children's school fees and meet their family's needs—something very few of their husbands attended to, having drunk their meager earnings at local bars. In time, Annie and Rose, both of whom had been disabled by polio and walked with crutches, joined us as cheerful salespersons.

Known and referred to as an "uplifting place," Uganda Crafts attracted a number of highly productive and talented artists. The late Jonathan Bifiramunda designed our logo of a woman weaving a basket, and along with others produced popular batiks for sale. Skilled in playing the traditional instruments he carved on location, Everisto worked on a mat near the entrance of the shop. James Mirembe, who was born without hands and had only stumps for legs, drew cards with the stubs of his wrists. The shoes and belts Christopher made could be ordered to size and were popular with customers. Very soon, other people with disabilities, as well as able-bodied artisans, supplied Uganda Crafts with a variety of colorful, useful and decorative arts and crafts. And contrary to the opinions of our critics, who said it was not possible for an NGO shop to be profitable, Uganda Crafts' sales substantially increased, with a profit of $50,000 the first year, $100,000 the second and early third, and $1,000,000 total sales for the three-year period.

Every Friday, it was Uganda Craft's custom to examine the baskets and other craft items dozens of women had brought, to select what we thought would sell. I then realized that the patio adjacent to the shop would be ideal for a coffee shop—for there were few places in Kampala where people could meet their friends over a good cup of coffee and a pastry. Betty agreed, and in just a few days, men built half-a-dozen round wooden tables and covered them with thatched umbrellas. We decorated the premises with pots of plants we had for sale. As chairperson of the International Women's Organization, which I had helped initiate, I requested the women who attended the meetings to bake cakes and pastries to serve at the coffee shop, which was growing popular. The women responded eagerly, including my friend, Joyce Choudhury, the charismatic wife of the Minister of Water and Land, who helped serve coffee and pots of tea to patrons.

It was soon obvious that we needed someone to supervise the coffee shop, so we hired Juliet Nalwanga, Betty's niece. Sweet of face and personality, Juliet was attractive to men and in time began to be troubled with a succession of ailments, before she died of Aids—the scourge that had begun to ravage the lives of so many of our friends.

Aids was first reported widely in Uganda in the 1980s as "slim." Nelson Sewamkombo and David Sserwada undertook the first clinical research to confirm HIV/Aids as the source of slim disease. Today Nelson is the founder of the College of Health Sciences, and he has helped reestablish Makerere University as one of the world's leading research centers.

The Aids virus did not discriminate, but found fertile ground among the rich and the poor, foreigners and locals alike. It left thousands of children without their parents. Casual relationships were common in Ugandan society. And although victims and their families wasted, what caused the illness was not yet known and mothers unwittingly infected the babies they left behind, who died from diarrhea, pneumonia, meningitis, or whatever else finally defeated their ruined immune systems.

The dreaded illness was spoken about with fear and hushed voices, and those infected by the virus were dealt with as pariahs. The cycle of death worsened with unhygienic traditional and clinical practices, with statistics that shocked the world. Now sessions we had with the women at Uganda Crafts Friday mornings to select their baskets included talk on how to and if it was at all possible to avoid contracting HIV/Aids.

Basutis, Bark Cloth and Traditions

The articles made from bark cloth the women brought to sell at Uganda Crafts varied in texture and shades of the beautiful warm earthy color. One day, while sitting at the coffee shop with the wife of Professor Kajubi, who taught at Makerere University, I asked her whether she knew about the process of making bark cloth.

"I do indeed," Elsie told me, disclosing that she had only recently completed her own research on the topic. "Many years ago," she told me, "Buganda women used to wear cloth made from the bark of trees known as the *suuka*, which draped under their armpits to their ankles. That was before early Arab and European traders introduced manufactured cloth. The inner bark cloth, or *nkutu* the women used to wear was tied around the waist by an *akeebajo*. This was covered with another piece that went round the back and under the armpits to cover their breasts and was tied by a wide piece known as *omebajo*." She rose from her chair to demonstrate how it was done. "The idea of letting the bark cloth sweep the ground has been there for a very long time—especially with those who have large round figures, which, by the way, were—and still are—a sign of beauty," Elsie said, smoothing her ample stomach with an amused chuckle. "But when the girls started going to school and practiced their physical drills, or while they were running to classes—even though the wrap exposed nice shoulders and left the hands free to work, the *suuka* had a tendency to slip down," Elsie continued. "So the clergy and teachers at Gayaza Mission High School encouraged the girls to add a square yoke with small pleats to the shoulders, which was fastened by two buttons with short sleeves. This became known as *kinagayaza* after the boarding school, or *bodinga*."

"Which was probably a mispronunciation of the English word 'boarding,'" I added with a laugh.

"And as you know," Elsie continued, "it's the style we still wear. It was then, about sixteen years after the turn of the twentieth century," Elsie guessed, "there was a tailor from Goa by the name of Gomesi who had come to East Africa with the 'Lunatic Express'—the rail line built by the British with the help of Indians they recruited and brought to Mombasa—who picked up the idea of making *basutis* for the women who lived around Mengo and Kampala. After that, our dress was also known as *gomezi* or *basuti*—'best suit,'" she laughed, clutching my hand. "But I've not told you how the bark cloth is made. The *omutuba* tree needs to be at least thirty-seven years old before the bark is stripped to make cloth. It is then spread over a log and beaten with wooden sticks until the width needed, after which it is wrapped in banana leaves and smoked over a pit of hot charcoal."

"Heating must create the variations in color," I surmised. The handbags, lampshades, place mats, hats and bed covers that we sold for popular sale at Uganda Crafts varied a lot from piece to piece.

Master Artist

Up the stairs, and deep in the corner of the former chief's house in Lugbara, stood the awesome figure Francis Xavier Nnaggenda had sculpted from *lusambya, omutuba,* palm wood and metal. Amazed at the sheer size of the figure and the impact it had on me, I backed up to take in the sculpture, which Sidney Kasfir says "clearly conveys violence and transformation."

Francis was a quietly dignified person, small of stature, but the figure he had wrought was imposing. It took time that afternoon for Francis, the caretaker of Francis' grandfather's house and me to load "Blood Rain Dust" into the back of the car. Fearful the ten-foot masterpiece would get damaged, as it was too long for the Land Rover, I drove slowly and carefully to Uganda Crafts, where we were delighted to have the incredible work of art dominate the display.

On a later day, I stepped carefully over half-finished carvings in the small dark room where Francis created his pieces. *Not again,* I thought, for I was not prepared for the immense emotion that swept over me. I marveled at the tragic, nine-foot, armless and headless sculpture that stood supported by one leaning leg in the center of the room. "Though it expresses the horror of war, the victim is a symbol of the tragedy and triumph of the human spirit," Francis said quietly, smoothing the beautiful wood with his hands. Then, chuckling softly he said. "I told the fallen *mukebe* tree I would give it another life and took the piece to interpret what was going on not only for Uganda . . . because when I was in Germany I saw what the Jews had endured," he said, his voice tapering off. Not long after, Cole approached David Court and the vice president of Rockefeller, who agreed to purchase "War Victim" for Makerere University. To this day, the powerful piece stands in the public library for all to see.

The Uganda Francis Nnaggenda knew had been a quiet and peaceful place before war ravished the beautiful land and destroyed its people and their culture. Born in the village of Bukimi in 1936, Francis had shown artistic talent from an early age. The Kabaka of Buganda recognized his talent and advised that he be awarded a scholarship. Francis left Uganda in 1963 and traveled to Basel, then Lucerne. With the help of a church scholarship, he entered the Academy of Fine Arts in Munich, where, after completing in three years a sculpture course that normally took five, he was honored with highest distinctions and designated a *Meisterschuler* in 1967.

For a time, Francis taught at the University of Nairobi in Kenya. "It was not easy for me to be a tutor for preliminary students under a conservative British system," he admitted. "I wanted to revive and instill heritage into young students, but the system gave me no chance. So I did pieces on my own and tried to create an artistic environment."

Identifying a large rock near what is referred to as Globe Cinema roundabout in Nairobi, Francis started carving. "And when I finished what I called 'Mother,' I talked to the university authorities and said I had completed a piece. But they were hesitant and said they would look into the matter." At the same time, Francis joined the Institute of African Studies at the museum in Nairobi and talked to Richard Leakey, who gave the university two weeks to respond to Francis' piece and if they didn't to let him know.

"Richard then called me," Francis continued, "and what he did was to call the Kenya National Army. I was told my stone was being removed with a crane. I ran down with my camera to photograph the drama, but the army didn't allow me to take a picture," Francis laughed, before adding: "So I was back in Uganda when I got the newspaper cuttings telling me the stone had been taken to the museum, where it stands to this day."

"Mother" is the first sculpture you see at the newly renovated museum in Nairobi.

Francis then traveled to France with his family, where they spent a year en route to the United States. While in Europe, he visited great cathedrals and works of art that he told me had "greatly influenced him and his image of God."

"But you cannot know what form He is," I said. "You cannot paint God. God is in everything—everywhere."

"For me, it is difficult to know why men destroy things," Francis said. "For if God created it, is it bad?"

"When you are walking, you protect your feet. But is the ground bad?"

"Should we burn the bush?" Francis replied.

Francis was awarded the Meritus Certificate by Who's Who in the World for his powerful expressionistic contributions to the field of artwork, contributions that Francis said "must possess social meaning to be incorporated wholly into the form that dissects, reassembles and enlarges the chosen theme." Although advised not to return to his country, which was at war, Francis returned to Uganda in 1978 "because I loved my homeland. My country was on fire." Known to be the greatest single influence on art in Uganda, Francis was hired by Makerere University to head the School of Fine Arts and the Department of Sculpture. He and his family lived at the house in Lugbara until security became an issue and they moved to the university apartments on campus.

"It was not easy to return to Uganda," Francis admitted one afternoon when we sat together on the verandah of the home he had built overlooking Lake Victoria. "When I drove there from Kenya there were many soldiers at the border. The car was like a moving forest, with nine suitcases strapped to the top. Life became so difficult. But for me to come back from the States, I realized how very right I was to share with my family the Uganda experience. In the days that followed, we had to sit down and find out how to get what we didn't have."

But what Francis lacked in material goods, he made up for with his magnificent works, and Betty and I were pleased when Francis agreed to become a member of the Uganda Crafts board.

Child-to-child

About the same time Betty and I opened the craft shop in 1983, Violet Mugisha, an enthusiastic primary school teacher and I pushed back the desks and chairs in one of the classrooms of Kitante Primary School and deliberated with the teachers how to set about starting a Child-to-child program. That meeting spawned a network that eventually spread to schools in more than ten districts of Uganda. We were of the opinion that children were capable of and should take an active part in their own general development and that of their younger brothers and sisters, and selected two pilot schools—one urban, one rural—where children were taught about preventive and curative health, hygiene, accident prevention and other caring activities appropriate to the local situation. The teachers, who were underpaid and often taught under demoralized administration in classrooms that had been looted of most essential materials, were actively involved and keen to promote the program.

In the months that followed, Violet and I approached the manager at the local radio station, who offered us prime time on Saturday mornings for the children to encourage the public to recognize they were indeed an integral part of their own welfare and that of their family and society. Thereafter, the children entertained Ugandan television viewers and translated Child-to-child objectives to other children through drama presentations, song and poetry. Encouraged by the enthusiastic response to Child-to-child, I compiled illustrations, songs and stories composed by the children in a booklet and on calendars I had published with financial help from the Norwegians through our psychologist friend, Magne Raundalen. The proceeds helped schools with materials and transport.

Six years after Violet Mugisha and I started Child-to-child at Kitante School, Cole and I returned to Uganda and looked her up. Full of confidence, Violet took my hand and led us to her office, located at the Institute for Teacher Education in Kyambogo.

"Don't forget, you left me with few staff and little of the support we had when Cole was the UNICEF Representative," she said, picking up the latest Child-to-child newsletter from her desk. "See," she said, pointing to the article. "Child-to-child Uganda won the Maurice Pate Award in 1991. We have really moved from strength to strength," she continued proudly. "We've expanded to over fifteen zones. Each zone encompasses four to nine schools that work in collaboration with the ministries."

Savoring my friend's enthusiasm, I smiled to myself as I reflected back to the last afternoon Violet and I had spent together, in July 1986, when her attitude and words were heavy with misgivings.

"Something presses on my heart," she had said tearfully. "I sincerely feel a very big load pursuing Child-to-child's aims without you."

Chapter XVI
Penny

MYRIADS OF TERMITES fluttered in the sultry air. Shedding their wings, the insects scurried frantically on the ground, only to be caught by small children who collected them by the handful to roast over a fire and eat.

Dark clouds threatened rain, casting shadows on the track Penny and I were taking to visit her sister, who I had never met. Tall grass brushed the sides of the car as I drove down the rutted hill to a rusty gate. Closing the gate behind us, the guard motioned me to park the car near a dilapidated building. Rather than entering the bare room of the mental hospital, we sat on a wooden bench on the small porch and waited. Neither of us spoke. Finally, a woman dressed in a shapeless faded pink smock shuffled toward us. Embracing Penny, she smothered her sobs in her sister's shoulder before turning to stare vacantly at me. Then, with a toothless grin, she came forward to let me kiss her on the cheek and take the food we had brought her.

Caught with feelings I needed to think about, I left the twins and walked along the barbed wire fence. Insects trembled in the wind. There was a terrible neglect about the overgrown premises. It was obvious that patients had to fend for themselves. Two men sat together in the tall grass and when one, who was dressed in torn shorts, took short, quick steps toward me, I walked rapidly back to where Penny was stroking her sister's thick tangled hair. So engrossed were they in each other's presence that I had to repeat my question as to whether they would like me to take a photograph of them together. Blinking back tears, I focused my camera on the twins and snapped the picture.

An ache of hopelessness filled my heart as Penny smoothed the frock over her sister's distended stomach. They embraced before we drove out the gate.

"I used to visit my sister often during the first three years she was institutionalized," Penny told me as she wiped away her tears. "She was so often sick and her head was thick with lice. At the time she didn't know me. They put her behind bars and kept her naked, as anything she wore she tore to shreds."

These words began Penny's story. Frequently having to stop to grieve over her recollected pain, I tapped out what she said on the keys of a portable Remington typewriter.

"My mother was a tall, attractive woman with long black hair and dark brown eyes," Penny continued. "She was born in England in the early 1920s. The army officer she met was killed in combat before their marriage date. She then began a friendship with her fiancé's best friend, John David Trendell, who loved her a great deal and who her parents advised her to marry.

Joanna and John left the shores of Britain in the 1940s and sailed to Kenya, where John taught at the Duke of York in Nairobi until he signed a contract in Western Uganda. Although stern, John was well liked and respected as the headmaster of Mbarara High School. The staff he recruited from the U.K. lived in a housing community within the premises of the school. John played the organ in the Anglican Church. Traveling occasionally, he pursued his professional activities while Joanna raised their three Uganda-born sons.

Mathias Tamuzaire studied at Mbarara High School. Blessed with good looks and a pleasant disposition, he earned his keep and school fees by cleaning the Trendell's house, planting their *shamba* and occasionally looking after their boys. Joanna was attracted to Mathias, and began an intimate relationship with him during one of Mr. Trendell's absences. She became pregnant and bore twins on 8 January 1962 at Mengo Mission Hospital in Kampala. As it was obvious the twin's father was African, Joanna's explanation to her husband was that she had been raped on her way to the outdoor privy.

The Trendell's sons attended boarding school in Nairobi and were never told of Patricia and Penny's birth. When the twins were about two weeks old, Joanna took the girls to London and placed them in Woodfield Babies Home.

"We were already thought of as *kaffirs* (mixed blood) at an early age," Penny recalled. "The older boys and girls liked us to do sexual acts. We did it for candy. By the age of six, I was a professional, oral mostly. One day I was taken forcefully and torn and needed to have stitches. I blacked out and was taken to the hospital. When I came back I could not tell on anyone . . ."

One day Pat fell from her bunk bed and received a head injury. When she returned from the hospital, the twins were moved to another home on Queen's Road where the children were accommodated ten to a room, each with a small metal bed and a bureau of drawers. "Miss Blair, the headmistress, and the nannies were very strict. If one of us wet our bed, we were spanked and locked in the bedroom all day without heating." However, Bonny, who Penny referred fondly to as "Auntie," became a close friend.

"Comfort was to go to her room in the evenings and she would make me a cup of hot milk. Sometimes I would wet my bed and Pat would tell Miss Suzanne about me. Anything I would do, Pat would report, and Miss Suzanne was merciless with me. We were not fed to our satisfaction, and one time I went to the kitchen and took a piece of cheese. Miss Suzanne found me and took me by the hair and threw me against a concrete ledge. My head was hurt and when I tried to tell Miss Suzanne it pained a great deal, she beat me. I woke up in some sort of clinic with a very bright light over my face. I screamed and then Miss Suzanne became very concerned and kind to me. But within months, she was back to the way she had been before. Claire, my best friend, and I seemed to be in disfavor more than the rest of the children and as a result we suffered, although we were, by then, only seven years old."

One day, Bonny called Penny and Pat to her room to tell them they were going to be adopted by someone who lived in Africa and that they would call this lady "Mummy."

"My immediate reaction to this news was that I didn't want to leave Bonny and Claire," Penny admitted. Penny refused to give Joanna a hug when she arrived and hid behind Bonny, who became stern and ordered her to say "hello." Keeping her distance, Penny cried desperately, and at first refused to get into the car to drive to the airport.

Pat seemed to have an instant rapport with Joanna and gave her a kiss on the cheek.

"Mummy chased me when I darted behind the plane we were boarding. When she caught me, she slapped me back to my senses. 'I hate you,' I cried, 'I don't want to go with you.' The stewardess sided with my mother and told us we were lucky to be going to a new home. I became airsick most of the trip and even had a temperature. Pat asked mother many questions the entire way, but I was quiet. Mummy worried about me. Our arrival in Nairobi was extremely uncomfortable. Our brothers' reaction was to get into their own car and drive away."

John Trendell drove Joanna, Pat and Penny to their large home in Karen, where they were frightened by the dogs patrolling the sprawling gardens. The girls were introduced to the servants and led up the stairs to a large room with a playroom attached, where they were told they would be taking their meals.

"'I beg your pardon, Mummy?' I would have to say. She taught us how to speak English properly, as we spoke with a cockney accent. She was impatient with me, as I was not quite as quick with my studies. At night, she tucked us into bed with a quick kiss on the cheek and I would often wake up to find Pat in my mother's bed. My comfort was my thumb and this became very much a habit even up until now when something traumatic happens. I was punished because of this; sometimes with plaster on my thumb, sometimes sent to my room and smacked."

"Was it because we were half African?" Penny wondered, as she and Pat were never introduced to visitors but were instructed to exit by the back door to go out and play at designated times. In time, the girls were invited to join the rest of the family for dinner, unless guests were invited, at which time they were given their food in the playroom.

The Trendell boys referred to the twins as *kaffir* or scum and resented the attention Joanna gave them. "They taught us bad words and were always doing something mean to us, such as making us fall off the motor scooter, then chasing us around until we fell again or putting us into a *siafu* (ants) nest. One of them was only mean to us when the other brothers were present. The rest of the time he would take us for rides and show us things. I remember he took us to see a giraffe. When I would spit words at them they hit me, and when Mummy would find out she would say 'boys will be boys.' For Pat's sake, I had to try harder because she could not handle the abuse.

"Mr. T never stepped into our room. He had nothing to do with us. We were to say, 'Hello, Daddy,' upon seeing him and that was our relationship. The only time he was called upon was when we did something wrong and he would be asked to spank us but he never did. He would just look at us with a certain look and walk away. I liked him and wanted to get closer to him but of course I could not."

Bewildered and angry at having to sit on the floor of the car rather than on the seat when Joanna drove them to school, one day Penny asked her mother, "Why? Even the dogs can look out of the windows, why can't we?"

"I was slapped for my insolent remark," Penny remembered sadly. "I cried the first three weeks mother took us to school. Pat and I were placed in separate classes and were discriminated against by the white children, who called us names and told us we must have been born in the streets with no mother or father."

After some time, Penny and Pat were transferred to Hospital Hill School, where both girls were proudly on the sports team. "But our friends were never allowed to come home and we were not allowed to go to our friends' homes," Penny recalled. "Mummy discouraged my friendship with a little French boy. When she saw him kiss me on both cheeks, she told me to play with girls. I enjoyed school very much and hated going-home time. I used to take my time getting my schoolbooks ready until mother got cross with me. I would then have to kneel at mother's side in the car and read to her. I never could pronounce colonel correctly and the other children called me 'thickhead.'"

"At home I played with the dogs. I was often sick and I enjoyed the attention the doctor gave me. Pat found her comfort with Mum. One time I was so very hurt when I wanted to help the two of them bake a cake but was told to go out and play. I wept about that and thought about Auntie Bonny. One time Mum told me to forget about Bonny, forget about the home, forget about the past . . . but I couldn't.

"About this time, an African man began to visit us."

Mr. Trendell had always liked "Tamu" (Mathias Tamuzaire, who had worked for them in Uganda) and even sometimes referred to him as "son." The girls adored him.

"We felt free with him and played and romped about with him. Our bedroom was very near to Mummy's room. I remember we had a very nice red bulb in our playroom and one evening when Mr. T was gone, the four of us were together. Pat fell asleep and I soon dozed off as well, but I woke up to see Mummy and Tamu fondling each other. So I pretended to sleep. They eventually realized they were in the wrong place to be behaving like that and embraced each other before walking out and leaving the light on. I was very surprised at my mother for doing this sort of thing, as we had always been taught that Africans were separate. We were even told not to have African friends. Somehow," Penny said reflectively, "it was then that I suspected Tamu as being my father. The next day, when Mummy was bathing me (Pat was always first), I asked her 'Mummy, how could you ever do such a thing with a black man? I saw you last night.' At first she kept quiet, but when she realized what I had said, she slapped me so hard I fell into the tub and saw stars. She then picked me up and warned me never, ever to say anything about having seen her. (I had once seen two dogs mating in the yard and asked mother what they were doing. She hit me with a stick and chased me up to my room where I was locked in without an explanation. I got furious and hit my head on the floor. I couldn't figure out what I had done wrong and what had happened to the dogs I saw together.) Another day I asked my mother why boys had an opening in the front of their pants. She answered with a question. 'What do you think it's for?' I returned with, 'If I had known what it was for, I wouldn't have asked.' She got very, very angry and slapped me. I went away sulking and again wondered what I had done wrong. I was often terrified of my mother. She was sharp with me and sometimes beat the hell out of me.

"Then suddenly Mummy began to spend a lot of time with me. Pat began to recognize the attention paid to me and began to tell on me, even if both of us had done the naughty thing. She was larger than me, and if there was a fight, she would win. Mummy continued to favor her opinion, but the brothers disliked Pat more than me and antagonized her."

Taking the girls aside one day, Joanna told them they would not be attending school on Monday. Although Penny had noticed their belongings being packed into a trunk, she never asked why, but wondered about it and had thoughts of missing Francois.

The boys became unusually kind and escorted Joanna and the twins to the train station. It was only while traveling that Joanna told the girls she, Mr. Trendell and the boys would be traveling to the U.K. and that the twins would be living with Tamu in Uganda.

"Certainly by this time, Mr T must have realized Tamu was the girls' father," I told Penny, who thought the same.

"We all burst into tears together," Pat remembered, looking off into the distance. Interrupting her thoughts, I asked Penny whether Tamu had been at the Kampala train station to meet them.

"Tamu had traveled to Mbarara," Penny replied. "But our auntie, Tamu's sister, who worked for the Ministry of Education, was waiting for us." Tearful and unhappy, the twins ate little of the lunch Phoebe had prepared. Requesting the twins and their auntie to accompany her to the station, Joanna told the girls that one day she would take them to London. For years they held on to that hope.

"Here at least you will be free," the girls' auntie told them. At first they found it difficult to adjust to a Ugandan household, having so often been told not to associate with Africans. Although they were eleven years old, they did not attend classes for some time.

Eventually the twins were sent to Kololo Primary School. Penny thought the mathematics teacher racist, as he was particularly cruel to the Ugandan children. Tired of him pinching her cheeks, hitting her with a ruler and pulling her hair, she skipped his classes and only appeared for tests. However, she performed well in English, as her teacher, Mrs. Lobo, was good to her. Pat, on the other hand, was liked by all her teachers. She was placed in class A, Penny in class B.

"I guess because I was uncomfortable with myself," Penny reflected. "I became a bully and frequently picked fights with the other girls." As a result, her aunt Phoebe was often called in by the headmaster to discuss Penny's misconduct. One day, Penny hit the monitor, who had rapped her on the head with a ruler. She was called into Mr. Khan the headmaster's office and told the next time she caused trouble she would be expelled.

"The monitor never touched me again," Penny recalled with satisfaction. "Pat kept asking me, 'What has gotten into you?' Phoebe tried to teach me how to wash dishes and clean, but I was clumsy about things. She got angry about this, as she valued her possessions. I reacted strongly to anyone calling me 'half-caste.' I would fly after anyone, bite, scratch and hit, battling someone or the other. But we became friends with the Norwegian girls who lived next door and who let us ride their bikes. We told them we had no mother and father. They used to give us pocket money and sweets, and Auntie became furious and said, 'Do you go to people's houses to beg?'"

On occasion, Tamu, who worked for the army, visited the girls. One day he invited them to spend two weeks with him and his lady in Mubende. Soon after the girls returned to school, Penny was told Pat had hit her head and was lying on the floor having a fit.

"I thought she was dying," Penny recalled. "The teachers brought an ambulance and we went to the hospital, where they checked her and we took her home. I advised my father to take Pat to see a mental doctor because a doctor friend told me her fit was a disease of the brain—so I assumed she needed a doctor. But he said, 'I can't believe Pat is mad.'"

Pat had no further problems for one year. Then one afternoon, when the twins were at home, "Pat looked as though she wanted to catch a mosquito."

"I knew something was wrong," Penny said thoughtfully, "when she gave a shout and fell on the floor. I jammed my fingers into her mouth to keep her from biting her tongue and she bit down. I screamed and Aunt Phoebe came running in wondering what was going on. She started hitting Pat and I shouted that Pat was not well. Finally, Pat released my fingers and they were bleeding. It took about half an hour for her to recover. Auntie took us both to the doctor—he explained to us that Pat had fits and gave her medication and me a shot for the bite. It was a matter of forcing her to take the medication—it upset all of us a great deal."

Sometime after the twins sat for their Primary 7 exams, Pat's boyfriend, Ben, who attended Kololo Secondary School, informed the girls it was not right they had not been told their birth mother was the white woman who had adopted them. Not really surprised, having seen Joanna and Tamu together, Penny was disappointed that Joanna had not sent the money and clothing she had promised.

"After that," Penny recalled. "I started asking myself, 'Why are we here, then?' I asked Ben a lot of questions whenever I could find him alone. He would answer with the little he knew. He told me that I was born by accident."

When Tamu became the commander of a paratroop in Lubira Barracks and was allocated a big house in Ntinda, Jinja Road, he moved the twins in with him. The Munyoro women he lived with knew of the girls' background. From the beginning she disliked Pat, who Tamu favored. Penny eventually picked up Lunyoro, the language Josephine abused them in.

Tamu suggested Penny withdraw from school for a year, but encouraged Pat to continue her classes at Gayaza High School. One night, when Josephine had gone to her village and Pat was home for the holidays, Tamu informed the girls he was their real father.

"From now on, you will be known as Tamuzaires," he told them, suggesting it was because of discrimination their mother could not keep them. Astonished the girls were not more affected, Tamu took them to the officer's mess to introduce them to his boss as his children.

The girls were delighted to have a father and were taken to court to sign various papers, but their birth certificates and records were never found.

Pat became hysterical soon after returning to school and was taken to Mulago Hospital, where Penny helped nurse her sister.

"Pat frightened me one night when she sat bolt upright in her hospital bed and called me 'Mummy,'" Penny remembered. "She was out of control and when I tried to help her, she clawed my face and shouted, 'Don't kill me, don't kill me,' before she lurched out of bed and stumbled toward the balcony. I thought she was going to jump from the sixth-floor balcony and held on to her legs screaming for help. Three men from the ward came in to the room and gave Pat a heavy sedative that made her groan. It was terrible."

Unable to walk, move or talk properly, Pat was given physical therapy, but the nurses and doctors were disturbed that she laughed and cried unnecessarily and talked of peculiar things.

"Even though she had bedsores and her hair was falling out, she was getting better," Penny recalled, "but we all realized she was not correct."

Dr. Billinghurst from the U.K. was surprised when Penny told him the two of them were John Trendell's children, as he had known their father. The doctor had been working in Mulago Hospital in Uganda for many years and thought the best treatment for Pat would be for her to visit her mother.

"Daddy (Tamu) then made plans with the government to transfer Pat to Queensland Mental Hospital in the U.K. At this point, Mother agreed. But at the last minute, she rejected the whole idea and plans flopped. Daddy got furious . . . the process had taken about a year. Mother was to have given half financial support, my father the other half."

Pat had eagerly anticipated the move to the U.K. When she found out she would not be going, she became depressed and tried to kill herself with an overdose of medicine. Penny found the tablets Pat had taken and called the nurses, who rushed her down to casualty and pumped her stomach. Penny was blamed for being irresponsible with the medication. Again Pat went into a coma and again she was put on a drip.

In time, Pat recovered and was discharged. By this time, Tamu had become impatient with Pat. Fortunately, the girls' Christian neighbors, the Musisi family, were understanding and sympathized with the girls.

"Daddy and Pat were now on bad terms. Pat was quite normal most of the time, but wanted everything her own way. To get her to take her medicine was a real struggle. She would hit me and at times refused to take it. One night I heard her getting out of bed. I felt her put something under my pillow. Then, when she thought I was asleep, she stabbed me on the arm and narrowly missed my chest. Daddy rushed in when he heard the commotion and threw her to the floor and she had another fit. After this, I moved out of her room. She once threw a kitchen knife at me. It barely missed me and went into the cupboard. She then hit me in the mouth and it started bleeding. When she saw blood, she said, 'What's wrong with you?' She didn't know what she was doing," Penny put her head on her arms. It took some moments before she was able to continue her story.

With time and care, Pat improved and was sent back to school. Mrs. Cox, the English headmistress, understood Pat's condition and took responsibility for her taking her medicine. Meanwhile, Penny attended Jinja High School, where she stayed with her father's Bunyoro friends.

"Mrs. Benon didn't like me from the start. I'd knock at the door after school and the neighbors would tell me that she was in but she took time to open the door. This went on day after day. When Captain Benon was away, she would only give me a cup of tea for breakfast and a very light supper. She never ate with me. I had no friends. I was miserable and cried a lot. I wrote to my father, but he didn't believe me and told me to finish school. I got so hungry and used to eat raw green mangoes for my lunch under the tree. One day I got very, very sick with diarrhea and vomiting. I woke Mrs. Benon, but she told me to wait until morning. I felt I was dying and had stools everywhere—on the floor, on the rug. The next morning, I crawled to the front door and hauled myself out and collapsed in the middle of the road. Someone found me and took me to Mulago Hospital. My father was called, and when he saw me he cried and I cried. I told him I didn't want to go back to Jinja. Then he understood and took me home to live with Josephine and him. I attended Kololo Senior Secondary School for the second term."

While Tamu was in Greece for military training, Pat climbed the stairs of the school chapel and jumped, despite Mrs. Cox's trying to persuade her otherwise. Penny was called to the headmaster's office and told Pat had been taken to Mulago Hospital. Finding Pat in plaster, Penny again moved into the hospital to nurse her sister, whose broken bones took time to heal. Despite having convulsions from time to time, Pat was eventually discharged. This time Penny returned to school and Pat stayed home.

"Pat became nasty and made sure something was amiss each day I came back from school," Penny recalled. "One day she tried to swallow stick pins. She had first broken the medicine cabinet and taken medicine. I found her lying on the floor. The wrappers and boxes told me the story. Saliva was coming out of her mouth. I ran to the hospital for an ambulance, crying all the way. Pat's eyes were turned up and she didn't seem to be breathing. They gave her oxygen and said she almost died. Some people even suggested I had put a spell over Pat."

The girls were again admitted to the same Mulago Hospital ward. Sadly, Tamu and Josephine stopped visiting them. Although the headmaster brought Penny her lessons, she no longer did well, as she found it difficult to concentrate. Becoming more and more depressed, Penny lost weight. She survived on tea, developing ulcers and taking tablets in order to sleep at night. Patients gave her money and cigarettes, though she was warned not to smoke so much. But coping was difficult. Ever watchful, Penny was frightened Pat would come out of her coma and try to kill herself.

Almost a year later, Pat came out of her coma and slowly learned to walk and eat again. The girls wept when Pat was released. The hospital had become their home.

"David and I became friendly on the tennis court. He was Ugandan, about 6'2" and good-looking," Penny said, remembering good times for a change. "We walked home together after school—I was about fifteen years old and very short. The kids called us 'Porsche' and 'Mustang.' Eventually, I told my father that I had a boyfriend and he told me he knew. David was scared when he met my father, because he feared army men. But my father liked him at first sight. David's stepmother was a Polish lady. We all got on very well and visited each other often. Pat remained at home with medication. She didn't like David and David was frightened of Pat—for he knew her as abnormal. I got to know when her fits would come and took her to the mental hospital for a checkup every two weeks. I felt miserable that she always had to stay home and talked about killing herself."

One day, when Penny arrived home from school, she found Pat hanging over the steps and rang for an ambulance. Pat's stomach was pumped out at the hospital. She and Pat moved back into the hospital. This time, Penny developed an eye infection and could hardly open her eyes to take her O-level exams. "I failed, I'm sure, because I couldn't see properly. David used to come and see me and bring me food. Daddy had given up. He never came to see us. More than five months went by. Pat remained in a coma for about three months. Again I helped other patients and went around doing errands. We were discharged, and soon after that, Pat and I were sent to Mbarara to be with my father, who was on leave. I didn't want to leave David behind, but there was no choice.

"In Mbarara we visited a witch doctor. He made us sit on mats and told us what was wrong with Pat. He said, 'She is being bewitched by your stepmother, and your father is going to be bewitched to hate both of you and Pat to hate you and kill you.' Daddy and I understood the language. Pat did not. "I became terrified of Pat. After this, the witch doctor gave us some revolting medicine. I had to chew a sort of root and then thrust it into Pat's nostrils with my mouth. Then I had to cook some root and wash her with it. All of this seemed to make her worse."

Tamu returned to Kampala, promising he would send money so the girls could join him. But the money never came. Penny was convinced her father's plan was to leave them in the village with their grandmother for good. They lived in a mud hut and learned to grind millet, cook *matoke* on three stones with the firewood they fetched and carry water from the well in jerry cans. Pat slowly became stronger, although she still suffered periodic convulsions.

"The problem with living in the village was the jiggers I got in my feet . . . and I missed David dreadfully. Sometimes he wrote and asked how things were going.

"Eventually we earned enough money with the vegetables we raised and hired a taxi to drive us to Kampala. We went to my father's house. He shouted at us and said, 'Who told you to come back here?'

"I said, 'So that was your plan—to keep us in the village to rot with no money. Well, we're back.' Pat called him a 'bloody bastard.' She tried to punch him on the face. My father got his pistol, but before he could shoot, I hit his hand up so the bullet went straight into the ceiling. I ran with the gun and threw it into the pond. He asked me what I had done with his gun. He slapped me once with the back of his hand and I couldn't open my mouth for days. Pat went at him like a wildcat and scratched and bit him. I was on the ground crying. Finally Pat ended up crying and the two of us were left alone.

"Daddy went out and came home drunk. He took another pistol

from his desk and after another quarrel with Pat, hit her on the side of the head with the gun. She went hysterical and ran onto Keera Road, shouting. My father and I ran after her and she hit her head on the ground. Some passersby came with police to help us, and we took Pat to Butabika Mental Hospital. They first had to give her a sedative with about ten people holding her. They then cut her hair and stitched her up. My father was with us all along and admitted to having hit her. He was in his uniform and they couldn't accuse him of anything. I hated him for what he had done.

"A friend of mine, Grace B, had a nursery school, and invited me to stay with her. My father agreed, which shocked me. I was free. I started teaching kids aged three to six—baby class—and enjoyed it. The work kept me occupied and gave me a little money for daily essentials. I also could visit Pat, who was sick in the mental hospital and at times didn't know me. She was behind bars much of the time. They couldn't feed her, and she became so ill that they again put her on intravenous feeding. During this time, the relationship between her and my father slowly improved.

"When the war started and the Tanzanians attacked Kampala, the mortars and gunfire terrified us and we moved into one room together. School stopped. Sometimes we would go across the valley through the swamp to Kibule for supplies, but fortunately we had a dry stock of peanuts, sugar and beans. No one ever bothered us, although the army men were killing people. During this year, we just lived from day to day, trying to stay alive. I saw nothing of Pat, and this upset me very much, for I knew she might be starving at the mental hospital. I was also worried about Dad, who was on the battlefield."

After the war, Penny moved in with the Musisis, who accepted her like a daughter. Unhappy that Pat was unwell in the mental hospital, Penny finally found out the whereabouts of her father, who had had been taken to Luzira Hospital when he was critically ill and put under arrest for his safety. Penny wept when she saw her father, who told her Josephine had vanished.

Realizing she needed to support herself, Penny went to work as a secretary at International Air Radio and at the same time attended classes at Uganda College of Commerce. David, who had been studying at Kololo Senior Secondary School, eventually went to Nairobi, where he completed his college course. He was unable to find a job when he returned to Uganda and Penny found herself supporting him.

"I even wrote to my mother in England for a loan, but she refused. Mrs. Musisi was curious as to why I was depressed and didn't eat properly. I changed toward everyone and was miserable, until I could take it no more. I came out with the truth and told her how badly I was in debt—about 140,000 shillings. She couldn't believe her ears. She told me she would pay the money and paid the whole of it with one payment. She told me to leave David alone. That he would ruin me and that I was encouraging him to be lazy. I realized my mistake and told him he must get a job, but he resented this. He said he knew what he was doing."

Problems arose at International Radio between the British manager, Mr. Benet, and the sales manager, Mr. Ochan from northern Uganda. Accusing his staff of being guerrillas, Ochan attempted to take over the office. Some believed it had been Ochan who held the grenade he said had been thrown at him. Having not been present when the others were picked up, Penny was the last one to be taken as a suspect at gunpoint to Nile Mansions. Mr. Benet escaped to Kenya.

"We were all beaten and asked who threw the grenade. No one knew anything, so they believed us and left us alone. Two days later, we were taken to the Central Police Station and kept in separate cells where no one could come and see us. Eventually, we were released on bail. David paid some 50,000 shillings, which he got from his mom, and was paid back eventually by Air Radio. We were then under surveillance but free. They told us to go back to work. David warned me not to return, but I said, 'Then what will I live on?' I went back to work.

"Ochan was in charge, and we were told to report to Central Police Station every month, until one day many of the staff fled to Nairobi without telling me and I gave up my job."

Meanwhile, David found employment as the assistant manager of a German company, raising the ire and jealousy of a woman from northern Uganda, who had worked for the company for six years.

"It was David's birthday, and he gave me money to buy and prepare food for his party. I promised to at Lubuwa Estate before 4 p.m. to cook. I completed everything, the visitors started coming and we waited for David. But he didn't come. I knew something was wrong.

"People started drinking and eating and were in high spirits, except for me. We went up the road to ring, but all the telephones were out of order. It was going on to 10 p.m., but David had not yet arrived. I finally had to announce that David would not be there. Everyone left except one person. Very early in the morning, we got into the car and drove to David's office. We found it locked and went to his boss's flat at Diamond Trust building. The houseboy appeared at the door. I asked him where David was and he told me that he had been arrested for having arms and had been taken to Nile Mansions.

"The Nile Mansions receptionist seemed familiar to me. He told me he had seen a tall boy taken up the stairs. I then ran to the police, which was not allowed. The police hit me and ordered me to sit down on the ground. They asked me to identify myself.

"By this time, I was frantic and told them my fiancé had been brought here and they asked me on what grounds. I said I didn't know. The police then pushed me into a room and I almost collapsed at the sight I saw. Men had been beaten until you couldn't recognize them. Some even had their eyes torn out. I couldn't see David among them. They asked me whether I was satisfied and told me to try Makinde. As I was walking out, David's boss saw me and shook his head. I thought he meant that David had been killed and I started to cry.

"He took me to his car and told me David was still alive but that someone had plotted against him. He took me straight to his flat and tried to soothe me, which took some time. The next day, I saw David with so many other suspects. He was on a verandah on the fifth floor, without a shirt. He was bleeding. I wanted to go and touch him but they refused, and again I started crying. I asked the man in charge for permission to take him for treatment. He asked me, 'How can you take a guerrilla for treatment?' We were sent away and went back to the flat. I rang Mrs. Musisi and told her the story and said I would be away for some days. I found David's parents crying at the flat . . . we had a rough time. Three days later, David was brought to the flat for treatment. A private doctor came to stitch his wounds. After he was treated, he was taken back to Nile Mansions to spend the night. His boss then negotiated with high officials for David to stay at the flat and report daily to Nile Mansions."

About a week later, Penny, David, Mr. Alpman and his wife were enjoying a glass of wine together, thankful that David was at last recuperating. There was a knock at the door. Men appeared from all directions, saying David and Penny were under arrest. Mr. Alpman swore at the men and insisted that they had no right to be there. Nevertheless, Penny and David were pushed and hit down six flights of stairs into the back of a Land Rover. They were told to sit on the floor at the feet of the military men. When they reached Nile Mansions, they were taken roughly up the stairs into a suite and again told to sit on the floor.

"Where are your combat uniforms?' they asked. "Where are your arms?" Asked all sorts of questions for which they had no answer, they were hit again and again until Mr. Alpman came in and pleaded with the officers to stop beating them. Penny's toes were badly injured. (They didn't see Mr. Alpman for another month, as he had been told not to interfere.) David, who was bleeding badly from the wound on his head, was taken into a separate room, where he was beaten and interrogated for more than two hours. Penny remained on the verandah until David returned, at which time three military men took her in for interrogation.

"They really beat me, and when I started screaming, they would beat me again and again, especially on the face, the knuckles, toes and back. When I would fall down, they would kick me in the back and shout to tell them the truth. I kept screaming, 'My God, my God, save me!'

"They said, 'Let your God save you.'"

This went on and on until Penny said she didn't care about anything anymore. Pushing a bayonet under her chin, they told her they would kill her. Finally, the captain said, quite suddenly, "Release her."

"I crawled out of the room crying, asking for David. A military man kicked me in the mouth and asked, 'What do you want with him?' I began to bleed from the mouth. I was then taken to where David was crying on a verandah. There were at least twenty of us—all bleeding, and when I started to talk to David, I got a lash and was told, 'This is no talking place.'"

"Penny, let's take a break," I said, sick with the horror of it all. "How could you endure so much?"

As if she hadn't heard, Penny continued: "The guards rotated . . . they would ask us what we had done and if we said we were arrested as guerrillas, they would hit us. We sat in the bare concrete day and night, about three yards apart, defecating and urinating where we sat. I was also having menstrual bleeding. The men all talked of suicide, but we weren't high enough to die if we jumped. Next door, people were tortured. Terrible screams filled the night. Nine of them were taken out while David and I were there. If people were taken to the godown at night, we knew they would die.

"We were full of terror, because we all knew our names would be called out. One night, David's name was called. I began to cry, and was told to shut up. I prayed for him not to be killed. I was praying with a doubt, but the Lord saw the position I was in. David was brought back after a long period of time. When I saw him, he nodded his head and I started breathing again. I cried bitterly and couldn't stop."

The days passed, and one afternoon David and Penny were told they would be taken to Makindye, where they thought they would be killed. Stinking and bleeding, they were led heavily down the steps to a Land Rover. When they reached the Central Police Station, they were hit up the stairs and taken to an isolated place, where Penny verbally abused her tormentors.

"I said, 'Shoot me.' David told me to shut up. We were finally able to see the criminal investigator in charge, who asked the other military, 'What have you done to these people?' He let us talk and was sympathetic, especially about me. His question was, 'What has this poor thing done?'

"After that, we were confined to separate cells full of water. I was alone. David was alone. The cell was small. There was two inches of filthy, stinking, cold water everywhere. I could hardly stand, as there was so much slime and excrement. He was badly injured and I encouraged him but he kept on beating and banging the doors. I thought David might become mad. Time was like a month. There was no ventilation. It was so windy. It was freezing cold. No one came to check. I couldn't feel my feet any longer. There was no food brought. David and

I kept shouting to each other.

"'Are you awake?'

"'Are you still there?'

"We wanted to be killed together. It seemed like days before a man opened the door and told us to get out.

"Could we move? I could not move. I was kneeling in the water. Numb. Cold—couldn't feel my fingers and feet. Crying with fear and pain. Someone came to help me into the women's cell. It was the size of a large loo. But it was better to be with others. David was taken to the male cell. I was given tea to drink—the best I've ever had! I asked for more. Nothing. I was shivering. The woman next to me, whose name was Janet, offered me her sweater. She started crying with me, and there were no words.

"It was dark when the night guards brought supper. They ordered the women to come for their meal. I crawled (Janet offered me a bag to put my food in) into the next room. I had stopped bleeding at least, but I was a stinking mess. We were given raw *posho* and beans with weevils. I was so hungry I ate it all, day after day, until I got such a bad stomachache I couldn't eat. I had diarrhea in the room. My head itched with lice. We ate, drank, slept . . . ate, drank, slept. Janet and I prayed together a great deal and got our strength from this.

"Another day, ten women were pushed into the room. Most of them had been badly raped, having been accused of cooking for the guerrillas. One woman was torn by a bayonet from her private parts up. I asked for water and tried to clean up the place. We took turns cleaning the cell. The wardens said, 'Sex or the razor blade.' I would refuse. Three of the women died. Janet and I had enough strength to soothe the women. There was a policeman on night duty. He especially harassed newcomers, by hitting them every time they spoke. One woman started crying and the police started kicking and hitting her so we warned those women not to cry.

"Three men came down the stairwell one night and called a man who had been in the men's ward for over two years. The man was questioned, beaten and shot through the head while we looked on. Today I cannot eat meat without remembering this. His head was blown apart and his brains spilled out everywhere."

Penny's tragic story weighed heavily on my heart, and after sharing her suffering and ordeal with Cole, we decided to ask her whether she would like to leave the country of Uganda and start anew in America.

She said yes, so we purchased a ticket for her to accompany Mother to Kansas, where Penny, at times disturbed, stayed in the small Midwestern town for a year and more before she returned to Uganda.

More than a year later, in 1994, I hardly recognized Penny. We had agreed to meet at Makerere University Guest House in Kampala, and I helped her heavily, awkwardly climb the stairs to a table under a tree noisy with turacos.

Penny took time to catch her breath. "How have your days been?" she asked me as I poured the tea.

"Well enough," I said. "And yours?"

"Really, my best experience was America," she began, with a smile that lit up her puffy face. "Thank you. In America I was too busy cleaning in the hospital to pay attention to my headaches. The people were friendly and good to me. But I missed Uganda and had so much pain thinking about Pat. I knew she had either died or was very sick, so I needed to come back." Shifting her body heavily, Penny sipped tea before she continued. "I was told Pat had been taken from the hospital. Dad had withdrawn all the money I had deposited and taken her to the village to a bloody goat shed. I was told by the youngest child that they tied her up. My father tried to starve Pat, but the children fed her. He used to rape her, molest her. He had become a beast, not a father," Penny said emphatically. "Pat finally died. A bottle was found near her. People said she had been poisoned with rat poison."

Please God, not more suffering, I thought, as I jotted down what she said.

"I had to go to the village to see whether it was true. My father was nowhere around, so I asked for the gravesite. I told them to dig up Pat's grave, and we found her in bark cloth. I bought a coffin, and Betty and I buried her properly with a priest and prayers.

"I came back to Kampala and was received back by UNICEF, to work as a data entry clerk," Penny continued. "Work was my only hope, and I could sometimes forget about Pat. In the meantime, I thought how I could kill my father. A year later, Father got a job in Entebbe in air cargo. So I went there to see him. When he saw me, he dashed away. So I went back to work. Next day, he came to where I work. My first reaction was to get the gun from the policeman. I grabbed the gun and cocked it and aimed it at him. The policeman tried to grab it. I said, 'He has come to kill me.' (He had a pistol at his side.) Father was sober and not even drunk. They radioed the police vehicle and took him to Luzira. I was so shattered, I had to take the day off. My friend Grace saw it.

"Betty and I went home together. I couldn't even cry. They said father escaped from prison and went back to the village. He started drinking crude *waragi* until he developed a liver problem. I was informed he was dying and thought, *Should I go or not*? I went with Betty and Charles Wogisha, who volunteered to drive with me. Father was very sick when we got there, so we drove him to Kitagata Hospital, where he was admitted and put on a drip.

"We were there for almost the whole morning. Finally, he opened his eyes and looked at me and said, 'Thank you.' He was sorry for what he had done. My friends said to say, 'I forgive you.' Maybe in my heart I said it, but not in my mind.

"I left money with my aunt to pay for his medication and to inform me when his condition worsened.

"Mathew Tamuzaire then died. I funded the burial. I was escorted by my friends. We had four cars. It was like any other funeral service at Sheema Kitagata, Buhenzi District, where we buried him on his land. They started naming the children who had put flowers on the coffin.

"I was called first. Then at least six other children came forward claiming him as father, ranging in age from twelve years to four months. They looked exactly like him. He had had many relationships. The people brought beer, danced and drank, sang and rejoiced. Sorry to say, we had such a good time. I was a new person. I felt so different and was relieved of fear. Father had hindered my happiness.

"I met my boyfriend in January, 1993. He was working at Hemu Ltd., a photocopying/bookbinding business. I used to say, 'I like that guy,' but how to say that to a Ugandan—a man like that. We admired each other. I used to say, 'You can't be a guy, what's wrong with you?' I told him we're moving to the new office and he asked me for a drink. I said, 'It's high time you asked me.' I couldn't concentrate and I bit my nails. We became close friends.

"The third of March 1993, I went for medicals for my job with UNICEF. The British doctor said, 'Penny, you have cancer and you are HIV positive.' I died inside. I sat and I looked at that doctor and told him he was lying. Before I could recover, he asked the next patient to come in.

"How I crossed the street to get into the car, I don't know. My friend Betty, who had lost two sisters and one brother to Aids, was not with me. I got home. No one was there. I locked the door and wondered what to do. I wanted to drive into a very big truck, bus or train. I thought how, how, how did I get this? David was alive, married with kids, healthy. I had had frequent bouts of malaria. Had I got it from the needles?

"Then—"

Covering Penny's hand with my own, I held tight while she continued her story.

"I battled with myself and thought I wanted to die. I cut my wrists, and almost blacked out. Then I looked at my arms and thought, *This is not going to kill me*, so I went for a knife. I held it, but got terrified of the knife (I had hated knives ever since Pat used to attack me during her tantrums). I thought about bashing my watch, taking the batteries. Finally, I just sat on the floor and wanted to die.

"I heard a knock on the door. It was my boyfriend. I said, 'Fuck off and leave me alone.' It was coming to evening and the neighbors came but left me alone. They called to me through the windows and got on my nerves. I found the car keys. Got in the car and parked near the Hotel Diplomat on a steep slope and thought I'd drive down and crash on the way down. I pushed my body under the wheel and went down a *murram* road, but no car came. I could hardly brake but used the handbrake until I stopped.

"I went home. My friends had gotten into the house and found blood. They called the police and I got back and locked myself in the loo. I shouted. 'Isn't anyone allowed to pee alone?' They made a hullabaloo until I opened the door and said, 'What is it?' I told them I was depressed.

"My friends said they would sleep with me in the next room. During that time, I had a battle with myself and I thought to see a counselor. I vacillated about seeing one, until finally one day I drove to Julie, who was an Aids counselor. She was just walking out to a funeral of one of her clients and said, 'I have no time now.'

"I said, 'If you don't see me, I'll kill myself.' Then I told her my story. She attended to me. 'I want another test,' I told her. She said, 'Are you ready for another test?' She had already taken a liking to me. I said I was ready to handle it this time. I got the result a few days later. I was positive.

"How to inform the people I cared about: my boyfriend, my personnel officer at work and Betty? I went to my boyfriend. I paced the sitting room for an hour, and then I told him, 'I expect you to walk from my life now and never come back.' He had tears in his eyes. I told him 'What I have to tell you is devastating. You should then leave my life.'

"So I told him I had cancer and was HIV positive. He told me, 'So what. It doesn't change the way we feel for each other. I will always be there.'

"I said, 'Thank you,' and asked him to leave.

"He did. I asked him to go for an Aids test. We went to find Julie, and he proved negative. I wanted to drop him. We went through counseling and agreed to remain friends. Up to this day, he sends me flowers and messages of love.

"We don't meet. I won't see him because I am human. He is human. We could make love. I even told him to find another girlfriend.

"He said, 'My girlfriend is Penny and will be Penny until the day I die.'

"On the 8th of January, 1994, I danced at my birthday party but didn't feel well. Soon after, I got meningitis and lost feeling from my hips downward. I couldn't see. But I refused to be hospitalized. They came home and put me on a drip. I could hear but not understand. Later I was taken off the drip and I was like a baby. A friend, Aggie, took care of me, and is taking care of me to this day. I was paraplegic.

They gave me crutches and started different therapy. That brought me back."

At this point, Penny asked for a piece of paper and drew an outline of a foot. On it, she indicated what the pressure points were for the heart, back, eyes and forehead.

"During meningitis," she continued. "I went into a coma. It was like an out-of-body experience in which I was battling between heaven and hell. There were beautiful people always in white with outstretched hands. They asked me, 'What do you have to say to come into the gates?'

"I would say, 'I am saved.'

"They said, 'Try again. Penny, you are not ready.' Then all of a sudden I was in a deep, deep place underwater. It was terrible. I saw faces and fire and the same people told me, 'This is where you belong.' Then I was tossed and thrown back to the beautiful people. Back and forth. It was like a year of battle. I saw red light. A very bright light. I could not see well. Like I had been traveling . . ."

Penny's tragic journey ended on 4 September, 1996. She died, attended by her friend Betty Mukibi, who wrote to me the following from Harare on 30 September, 1996.

"I should mention that Penny matured into a highly spiritual girl until the very day she passed away . . . I witnessed her very close, very sincere, devoted and true love for the Lord and I believe this was the single most important source of the incredible will that sustained both of us through the very trying times."

Chapter XVII
The Coup

THE CHILDREN WERE licking the syrup from their breakfast plates one Saturday morning in July 1985, when Jhorna reminded me to turn on the radio to hear the Child-to-child program. Turning up the volume, I joined the kids in singing enthusiastically:

"Child-to-child for the children of the world, children of the world,
children of the world, children of the world.
C for children like you and me.
H for health and hygiene,
I for you and you and me.
L for learning, and
D for doing helpfully."

It was Jessica's turn to bring the breakfast plates to the sink. While we were doing the washing up, the radio broadcast was abruptly interrupted by Tito Okello's voice.

"Citizens of Uganda," he began, "you are advised that the evil forces of Obote's corrupt regime have been eradicated."

Stunned, I hurried to find Cole, who had just received a call from the UN Security Advisor, Bob Robbilico, asking him to attend an urgent Heads of Agency meeting.

Short of breath from running up the stairs to the top floor of the IPS building, Cole found a chair among other heads of agencies gathered at the UNDP office to listen to the security advisor.

"Troop movements have been picked up by the American satellite surveillance. They are coming down the Bombo Road and are expected to arrive in Kampala by noon, so there should be enough time to alert your staff."

No sooner had Robbilico completed his directives than the sound of military machines rumbled below, and a convoy of fifty trucks, filled with troops from the north, rolled down the street. Cole called me and was just saying, "Everything seems to be fairly orderly—" when a blast shattered the window he stood near and bursts of gunfire interrupted his words.

The Uganda National Liberation Army (UNLA) was jubilant about their easy entry into Kampala. Clutching their firearms, they leaped from trucks to shoot or kick down the doors of shop-owners, who were all too familiar with looting and settling old scores. Those caught unawares fled or crouched behind buildings until they too joined the forces to strip the shelves bare. Staggering from abandoned stores with commodities, mattresses, appliances and whatever they could carry on their heads or tuck under their arms, the looters fled while artillery shells whistled through air hazy with gun-smoke.

"As for me, I wanted to loot more," a young friend, who was wise to the rules of the streets, told me some days later at Nakasero Market, where I had gone to purchase much-needed produce. "People were running and soldiers were shooting into the air," he said, pushing a large basket of vegetables onto the back seat of the car. Slamming the door with a flourish, he told me he had taken live chickens, fruit, money and even the weighing scales. "That day we were taking what we wanted," he said with a wide grin. "We took from others and they looted from us. We lost our bicycles. Then they started shooting bazookas and guns and we got frightened, but wanted to loot more. I wanted a motorcycle and clothing from the big shops. The soldiers took some of things, but not all. I saw two people who they shot because the soldiers wanted their jackets. I saw one of them who was dying. That man looked up at me and said, 'Where I am going, you will also follow me there.'"

Here James paused, dropped to the ground and traced circles in the dirt, reflecting on what he had told me.

Less than a fortnight later, Emmanuel leaned his bicycle against the side of a shed at the same market. Pensively peeling the banana I gave him, he told me James had stolen goods for the last time.

"He was beaten to death," he said.

Dismayed, I heard his explanation. The community James had been part of had had enough of his thieving habits and battered him until a final blow silenced his last words: "I'm sorry."

Against the advice of others, following an hour of frenzied shooting and looting, Cole ran down the stairs of the IPS building two and three at a time to reach the street. Radioing upstairs to the security officer to say he was leaving, he draped a light-blue UN flag over the bonnet of his 1957 MGA and drove to the UNICEF office on Bat Valley Road. "Leave for your homes as soon as possible, in whatever way possible," he advised his anxious staff. Locking the doors of the office, he and three others braved the anarchy of the streets.

"It's madness out there, sheer madness!" Cole told me when he arrived home. "Vehicles were being overturned, some of them burning. People were fleeing. Everywhere, people were running from the Acholi soldiers who were hell-bent on retaliation—settling old scores," he said grimly, as he parked the MG inside the shed and padlocked the door.

Fearful and excited, our children followed their father up the stairs to the verandah, where they fidgeted near him while he radioed his staff in Kampala, those up-country in Soroti and the Nairobi Regional Office. Throughout that unsettled afternoon, UN security reported undisciplined troop behavior and warned us to exercise extreme caution with our movements and to adhere strictly to the dawn-to-dusk curfew. Evacuation plans were mobilized for nonessential UN and embassy staff and any other foreigners who wished to leave. Fifty vehicles drove in convoy to the border, but the children and I had no intention of leaving.

Adding to our concern, the normally erratic electric power dimmed before it went off. Soon the *grra-grra-grra* of the generator added to the by-now thirty hours of uninterrupted gunfire on another dark terrifying night. Gathering ourselves into a close bond of reassurance, we read of Elsa's antics in *Born Free* and stories of trust and faith from the Bible. Each of us then prayed earnestly for God's protection and care for those traveling through deep sorrow—and for our own strength to cope.

How does one even attempt to clarify the madness of war: conflict, enmity, hostility—to clear, clean minds? I thought. Hate raged that dreadful night, in direct contradiction to what we had tried to teach our children regarding the fruits of the spirit: love and forgiveness. The words caught in my throat and tears stung my eyes when I tried to join Cole in singing "Michael Row the Boat Ashore," and I was only able to join him with "Jordan's river is chilly and cold . . . chills the body, but not the soul . . . ," all the while thinking, *Despite the conflict of things—all human beings have a spiritual soul.*

Drained from the uncertainty and trauma of it all, Cole and I exchanged hugs with the children, and told them, "God bless you, sleep tight, I love you." With no metal grilles to unlock, we walked through the French doors that led out to the large verandah into the luminous night. "Do you smell the queen of the night?" I had just exclaimed to Cole, when bullets hit the ground in front of us and we hastily left our chairs.

Cole grabbed the hand-held radio and reported, "Heavy popcorn close at hand."

Holding the walkie-talkie close to my ear, he asked me to decipher the barely audible whisper, which I thought said, "Uncle, Uncle, Uncle (Cole's code name) . . . do you read me?" Fairly sure it was the voice of the wife of the expatriate UNICEF employee who lived five hundred meters across the hill, I told Cole, "I think she's saying, 'The man has a gun and he says he will kill . . .'"

Informing security over the walkie-talkie that he was going up the hill, Cole shouted for me to stay with him on the two-way radio, then jumped into the Land Rover and coasted down the drive.

Tense with fear, I gripped the mouthpiece of the radio as if it were a life-cord with Cole. The children, who had been roused from their beds, sensed my apprehension and stood nervously, supportively near. *He must have reached Blankert's gate by now*, I thought. *He had to have gotten past Mbuya military barracks red-hot with gunfire—most likely the soldiers were reluctant to stop the Land Rover, fearing a senior officer might be driving it—who else would be crazy enough to be out?*

Cole gunned his engine through Blankert's open gate. Flashing his lights, he laid on the horn and slowly drove down the driveway. Six or eight armed thugs had overpowered the two police guards, and when they began to loot the house, Blankert's wife and children locked themselves into a back bedroom and whispered pleas for help to Cole's code name, "Uncle," over the walkie-talkie, while Bill dealt with the abusive and threatening robbers. Cole watched the armed robbers flee across the front lawn and left the vehicle running. Hurriedly helping the terrified family into the back of the Land Rover, he covered them with a mattress in case he would be stopped when he returned home past the volatile Mbuya barracks. Acutely aware that celebrating soldiers were drunk on victory and local brew, he drove back down around the smoldering barracks—a distance of no more than a kilometer or two. The children and I were ecstatic to hear the vehicle and ran down the stairs to welcome the distraught victims of terror.

Only then did we turn our admiring attention to Cole, who stretched out his shaking palms to show us the blisters that had formed from gripping the steering wheel.

What a hero . . . but how high the risk!

Again it was the children who brought some degree of normality to that troubled evening. Anxious to make our frightened guests feel comfortable, they helped me make the beds and put food on the table. As we were setting out the plates, Jessica asked, "Mama have you heard anything from Kuku?"

"I'll try to find out about her in the morning," I said, remembering someone had told us one of Obote's young sons was still at State House.

Tea and Terror

The coup d'état as told by Kuku Barua.

"It was a dark day in 1985," Kuku said softly, her dark eyes behind spectacles meeting mine. "Could I ever forget those Obote years? Despite having met so many heads of state, I have never experienced what I came to know in Uganda."

Kuku rearranged the shimmering sari over her shoulder and began her story:

"Heavy clouds promised showers. Mama Miria was selecting what she wanted to wear at the Women's Conference she was to attend in Nairobi. And while she handed me the clothing she wanted packed in her suitcase, she warned me the situation in Uganda was not at all good."

"Be careful and attend to Mzee . . ." Kuku remembered Miria telling her as she watched the President's jet fly the First Lady through threatening skies toward Kenya.

Soon after, whispered rumors confirmed Miria's words.

"Handle the china carefully," Kuku instructed the men, who were filling the boxes with what was needed to feed the dignitaries. Despite recent disturbances, the President had requested Kuku to accompany his entourage to the border town of Mbale, where President Daniel arap Moi from Kenya joined them for lunch, before Obote drove on to Tororo. "It was the last occasion on which I was to watch Obote act as the President of Uganda," Kuku remarked thoughtfully

Some days later, Kuku, who had been sitting on the verandah of her cottage near State House, heard loud, curdling screams coming from the guard's residence. Alarmed, she inquired of the staff what had happened, and was told four Acholi rivals had been castrated before they were killed.

Distraught and frightened about the obvious signs of insecurity and what she had heard, Kuku packed a suitcase with clothing. From then on, she shuttled back and forth between Entebbe and Nile Mansions in Kampala.

"I'll never forget that day, the 27th of July," Kuku said, shaking her head. "My daughter's words were barely audible on the crackling telephone line from India. She was shouting, and kept repeating, 'Are you all right?'"

"We're fine." Kuku responded. "Obote's little son Benjamin and five of his friends have just taken something to eat."

"But haven't you had a coup?" Kuku's daughter inquired incredulously.

After quelling her daughter's fears, Kuku put down the receiver and looked anxiously out the window to see security men in battle dress running in every direction.

Hurrying out to her verandah, she summoned one of the fleeing guards, who stopped for a moment to tell her that Tito Okello's troops had taken Luwero the previous night. "Kampala has been taken!" he added.

Kuku later heard that President Obote had left his home in Kampala and fled to Kenya by road with a convoy of five cars and a Land Rover. He was seen at the Busia border post, seated between two other men in the back of a dusty Mercedes-Benz. Kenya's *Sunday Standard* reported and that he had "apparently having been driven overnight to avoid both Okello's soldiers and the main roads. He and his 22-man entourage checked in at the Busia Teacher's Hotel, incurring a bill of Shs 100,000 which has yet to be settled."

In a much later article in the *Daily Nation*, Andrew Mwenda interviewed the former president and quoted Obote as saying his convoy arrived in the border town of Busia about 6 a.m. Soldiers who had heard of the coup tried to block their way, but others allowed them into Kenya. Following breakfast with Kenya's President Moi, Obote drove to Nairobi and his friend Kitili Mwendwa's house, where he found Mama Miria distraught about Ben.

It had been years since our friendship with Kuku Barua began in the shaded garden of our mutual Ethiopian friend and Cole's colleague, Araya Assefa, a man who always lived life to the full. Slaughtering a goat for the occasion, Araya moved about the lawn of his residence in Kampala with a tray of meat, encouraging his guests to enjoy what he had roasted. Declining his offer, I reminded Araya, "You know my preference for vegetables."

Flashing his gold-toothed smile, he replied. "The Brahmins in the south of India must have influenced your taste," before he introduced me to the attractive Indian lady who stood near me.

"When were you in India?" she inquired politely, selecting a small bony piece of meat on which to nibble. Starting up a conversation, Kuku told me she had been born into a renowned family from Assam "many years ago." She led the way to a table set with tea.

"I was born in Calcutta in 1927, the second of three children," she said, twisting the sparkling rings she wore on her fingers.

Kuku was schooled in convents before she married and raised her three daughters. At first, she accompanied her husband from post to post, but then accepted a job working for the Prime Minister in India's capital. Her responsibilities at the Rashtrapati Bhavan, the home of Viceroys, included the supervision of housekeeping, comforts and entertainment for the late Indira Gandhi's visiting dignitaries. Five years after Kuku had relinquished her duties at the Bangabhavan in New Delhi in 1982, Uganda's President Milton Obote conferred with the foreign office of India and requested Mrs. Barua's services.

"Obote walked into the room with only stockings on his feet the first time I met him," Kuku told me with a smile, her natural grace enhanced by the years she spent with the privileged of the world. Conscious that she had been the only one to stand up when the President entered the room, Kuku was amused to hear the secretaries refer to him as "Mzee."

"Obote motioned for me to take a seat, and inquired whether I liked Uganda," Kuku recalled. "I told him it would take time for me to know whether I liked the country. He chuckled amicably enough," she remembered.

He then suggested that Kuku fly to England to meet with his wife, who would know more about what would be expected of her at State House.

"When I arrived in London, I was taken straight to the hair salon where the First Lady was having her hair done," Kuku told me. "She greeted me warmly, and over a cup of tea told me what my responsibilities would entail." Charmed with Miria's sincere approach, Kuku accepted the First Lady's invitation to join her for dinner. "Our conversation lasted well into the night," Kuku said. "When I excused myself, Miria inquired when I would join them in Uganda."

Kuku returned to India in October 1982 to tidy up her legal matters and inform her husband, Brigadier Barua, that he would also be employed if he accompanied her to Uganda. "To Africa," Kuku told me quietly. "To a land and a people I knew so little about."

Pouring herself a cup of tea, Kuku sat back in her chair and continued her story:

"I was offered a cottage near Uganda's State House in which to live. The view from the verandah was of Lake Victoria, and so peaceful . . . so I chose to stay there, although I also could have taken rooms at the Lake Victoria Hotel in Entebbe."

The cottage Kuku resided in was close to the President's official residence, which Tanzanian soldiers had looted when they invaded Uganda to overthrow Idi Amin in 1979. Taking the items they fancied, the soldiers ignored the books and pictures, one of which was a Rembrandt print Kuku later presented to my mother. Amin's bedroom was littered with "pornographic magazines, vast quantities of medicines and State Research Bureau files on alleged subversives," as Avirgan and Honey wrote in *War in Uganda*. Every room of the house, including the children's bedrooms, had "caches of weapons, including submachine guns, pistols and hand grenades." In one of the bathrooms was an assassin's gun and a rifle with telescopic sights and silencer, plastic explosives, and "on a bed the British Admiral's hat that Amin liked to wear." Finally, the Tanzanians looted the State House storerooms ". . . stacked to the ceiling with transistor radios, watches, electric kettles, linen sheets and television sets"—goods brought from Stansted, England, on Amin's infamous "whiskey shuttles," which were intended as perks for his soldiers.

"For sure, alcoholic drinks remained a priority in State House," Kuku told me with a chuckle. Soon after taking over her duties, she was asked to take stock of crates and crates of beverages, along with boxes of linen, china and cutlery. While she was busy sorting through invoices, plumbers and carpenters from the U.K. supervised the installation of new bathroom and kitchen fixtures. Pure Thai silk was stretched over the walls, crystal chandeliers were brought in from the U.K. and furnishings from Denmark and Belgium were installed.

"Mama Miria warned me that thieving was not only a problem but a habit and that things would disappear," Kuku remarked, shaking her head. "I felt a real fear about the matter. I was so uneasy that when men carried boxes of imported whiskey and wine to the storeroom, I stationed someone at the foot of the stairs and I stood at the top. But even then, the bottles were pilfered," she said, still incredulous. "Could the men have stuffed the bottles into their trousers?"

Tipped off not to stack cartons near transoms in storerooms, in later months Kuku found that keeping the keys to the liquor store was both an advantage and a headache.

"There were endless requests made from those in both high and low positions. I always told the officers to put their requests in writing.

"It was February of 1983 when Mama Miria moved into State House along with the youngest of her sons and a couple of his friends. She called me 'Mama Kuku' and we became companions.

"The little boy, Benjamin, was shy. He was quite different from his brothers, one of whom was very black like his father. When not attending public school in Kampala, he spent most of his time with me. He was a stubborn boy, very fond of his mother, and he clung to her whenever he could. After he and his playmates finished riding their bicycles, they jumped into the bathtub together. They were not at all inhibited," Kuku laughed.

Convinced of the value of their companionship, and committed to earning their confidence, Kuku made friends with the State House guards. "I often sat with them when they were ill or if one of their relatives died," she related, with obvious caring. "They and the staff who worked at the President's residence were down-to-earth people. Very poor. They were very helpful with helping me organize the many dinner parties and preparing the guest rooms for the likes of President Kaunda from Zambia."

Low clouds lay over Lake Victoria the day Joseph, with UNICEF flag flying, chauffeured Cole and me under the arched entrance of State House. Resplendent in a gorgeous silk sari, Kuku stood with the First Lady in the foyer to welcome the ambassadors, heads of agencies, dignitaries and their wives to the mid-afternoon function. Directing Cole to where the men were congregated, she took my hand and escorted me into an elegant room walled in pastel shades of silk, where I shared pleasantries with ladies dressed in their national costumes. In time, we were led through the dining room past a very long polished table flanked by high-backed chairs, onto the back verandah, which was fringed with pots of flowering plants. Although the grass was wet from recent rain, the sun soon burned through the vapor and dried the lawn.

The musicians, dressed in their party color, red, set up their instruments and equipment before they struck up a catchy tune. Responding to the provocative beat, a long line of Uganda People's Congress (UPC) supporters swayed their hips and shuffled their feet. One woman pushed the bottle of beer she had been drinking under her chair and grabbed my arm to accompany her toward the animated group. Smoke curled from the charcoal fire and goat kebabs hissed over hot coals. Partiers roamed about, chewing the skewered meat. Women clustered together in their finery drew me into their circle and now and then I caught a glimpse of Kuku, who was preoccupied with her numerous duties.

"Many of those functions lasted all night," Kuku told me with a sigh, recalling that it was not safe to drive back to Kampala from Entebbe in the hours of darkness. "Most ministers would bring their wives, but sometimes other women accompanied the men for much wine and song. Normally two bands performed for occasions. Mzee preferred to dance on the verandah. But Mama liked to move out on the grass and every now and then, they would put down more liquor and more. We never ran out of beer. Our supply was limitless."

Though he normally commuted between Nile Mansions at the International Conference Centre in central Kampala and his Kololo residence, the President occasionally spent the night at State House with his family. One day, Obote walked into the room to find his youngest son seated on Kuku's lap playfully pulling her nose. Rising respectfully, Kuku greeted the President. She was interrupted by Benjamin, however, who reminded her not "to be afeared of my father."

"When I lost my husband, the President took my arm and wept," Kuku recalled sadly. "He was very sympathetic at my loss and suggested we not proceed with the Christmas lunch we had planned for many people—so only a few friends and family members joined the Obotes that day." Then she remembered, "Oh yes, one of the guests was the Commander of Obote's Uganda National Liberation Army, Tito Okello . . ."

"It was midmorning on 27 July when the shooting started in earnest," Kuku recalled. "Staff were running. All those," she said, managing to laugh, "except those of us left in the dining room."

"You had better eat your chicken and rice," Kuku advised the children.

"We're not hungry, it's too early," they complained. But Kuku coaxed them to finish their food. "There has been trouble," she told them. "We cannot be sure where the next meal will come from." But the boys ignored her concern. Leaving the table, they ran outside pretending to shoot at the enemy whose bullets now rent the air. Kuku followed the children and shouted at the guards standing by: "What are you waiting for? Leave your guns. Run."

"And run they did," Kuku said, remembering the madness. "They ran in their civvies after they ripped off their uniforms in flight."

Aware that the State House driver might leave the hill with the car, Kuku dropped the keys of the President's limousine in her pocket. Hurrying to her cottage to grab clothing and her radio, she dialed her friends in Kampala, who, she recalled unhappily, "would not hear of protecting me as long as I had the President's son."

"Take care of yourself," they warned.

Frantic with concern, Kuku requested one of the senior security men who lived in Entebbe to shelter her and the children.

"I will try to help you," she was relieved to hear him say.

"I could smell the gunfire when we drove down the hill to the house of a man who worked with the Fishery Department on the banks of Lake Victoria. But when we arrived, he did not want the official car parked near his residence."

"Then take the car and lock it in one of the State House garages," Kuku told the driver. But this did not allay their host's fears, and Kuku was distressed to hear him say,

"Please take the children. Move on to some other place."

"Where shall we go?" Kuku cried, with real concern. Desperate, she phoned a woman she knew in Entebbe, who agreed the children and Kuku could stay with her. Kuku then sent the driver with the car back to State House and she and the children trekked for three-quarters of a mile along the road, with trucks and cars careening past.

Confused and tearful about the traumatic change of events, Benjamin demanded to know why his father was no longer President.

"He cried for his rubber snake and the miniature car he left behind," Kuku recalled sympathetically. "He was just a little victim of unfortunate circumstances. That night he slept with me, and the rest of the boys slept two to a bed. Naturally, they felt insecure and some of them wet their beds. I spent my time washing sheets and kaftans. I felt so uneasy when one of the boys became very ill with a fever of 105, but did what I could with cold compresses.

"On 29 July, the boys watched a video and we celebrated Benjamin's birthday with dry bread and peanuts. The next day," Kuku remembered, "our host found meat at the local church, which the children devoured ravenously."

The boys became increasingly restless within the confines of their restricted quarters. Fretful, they complained of hunger when the food stocks ran low. Unable to purchase food from the nearby small shops as the *dukas* had been looted, one day, during interval lull in the shooting, Kuku walked up the hill to State House, accompanied by her hostess's guard, to where she knew she could find provisions to last the crowded household a few more days.

"On the third, or was it the fourth day?" Kuku said, trying to recall, "I rang Mama Miria in Nairobi to tell her, 'Your son is safe.'"

"Miria was emotional and said, 'Thank God you're safe, I had given up hope. Where are you?'"

But Kuku was reluctant to tell her of their whereabouts and repeated, "Benjamin and the others are safe and well. Please don't worry." Replacing the receiver, Kuku then rang the Indian High Commission in Kampala and requested them to inform her people in India that she was all right.

"Forget the child," the official told the astonished Kuku. "Where are you? We'll send a Corps Diplomatic car for you. We have been receiving so many inquiries about you from Delhi."

"I'm sorry," Kuku replied. "I can't tell you where I am and I will most certainly not leave the children," and slammed down the receiver.

"We had been in hiding for about six days," Kuku recalled, "when I heard that the officers in Entebbe were combing the area for me." Taking a deep breath, Kuku smiled and said, "Guess what they wanted? They wanted the liquor key! When I was making my way down the street toward State House," she said, "the new Commanding Officer picked me up and drove me the rest of the way."

"The car you have been using has many bullet holes and has been vandalized," he told her. "But don't worry: you will soon have a new Mercedes."

"I found only the TVs, fans and VCRs missing when I checked the rooms of State House," Kuku remembered, then surmised: "And those things were probably taken by the house staff. The terrible thing I had to deal with was those poor Lango guards . . . how they were butchered."

Although her bedroom closets had been opened and her clothing strewn around the floor, Kuku was relieved to find her cottage had not been looted. Subdued and anxious, she returned to where the children were to announce to her host that it was relatively safe to travel by road.

With typical Ethiopian caution, Araya Assefa assured Kuku he would help her evacuate the President's son. Later, in Cole's office, he lowered his voice and told Cole that Kuku had Obote's son, and the two of them hatched a plan.

The boys would join the convoy organized to evacuate the international community from Uganda at Lugogo bypass at 10 a.m. Designated as the UNICEF Security Officer, with a Land Rover equipped with a high frequency radio, Araya led the procession, in which were many UNICEF staff and their dependants. (Although we were told that as a family we were risking our lives, the children and I signed a waiver saying the UN was not responsible for our safety and chose to remain with Cole in Kampala.)

That evening, Araya arranged for the convoy of vehicles to park on the grounds of a secondary school just west of the border town of Busia. Anxious to keep the atmosphere easy, Araya purchased several sheep, which he roasted to feed the local army, those traveling and the police he requested protect the camp as the border was closed because of the coup in Uganda. While the foreigners found sleeping positions uncomfortably in their cars, Araya and the boys spread a tarpaulin near the fire and slept under African skies.

The next morning, Araya led the convoy through the Kenya border post. Greeting the immigration officials with his gold-toothed smile, he strode about the premises with Obote's son and friends following him like puppies. When asked by the immigration authorities about the boys, Araya told them they were his escorts and he would return with them to Kampala. (It was not uncommon for soldiers to be accompanied by young boys who looked up to them.)

On the Kenyan side, Araya left the convoy and drove to the Caltex petrol station 100 meters across the border, where he handed the boys over to an envoy that President Obote had sent to collect them.

Meanwhile, Kuku bowed to her family and friends' pressure and wrote to the Head of the President's Secretariat to request permission to leave the country. "But I was told the new President required my services and would like to see me."

The next day, she dutifully reported to Nile Mansions, where she waited for some time for Tito Okello to arrive.

"I could smell the cloth of his new uniform when he finally entered the room," Kuku recalled. "Okello beamed at me and pointed to a chair, on which his hat lay upside-down. So I picked up the hat and sat down and said I wanted to leave Uganda."

"You will be well looked after if you stay," the President told Kuku in broken English.

"But I really need time away," Kuku insisted.

Convinced of her concern, Okello consented for her to take one month's leave. Wishing her well, he stood up to shake her hand.

Curious to see the state of the room Obote had occupied, Kuku walked down the hall of Nile Mansions to find the room stripped of furniture, and the music system and video cassettes missing. "But I do remember that rows of books still remained on the shelves," she told me with a chuckle.

Much later, when Kuku paid Obote a visit in exile, she related the incident to him and was amused to hear his reaction.

"Those people don't require books," the former President exclaimed. "The books will still be there after ten years, simply because they do not know how to read English."

Before the third and last evacuation convoy was scheduled to leave Uganda, Kuku spent the night at the Speke Hotel in Kampala. As she was preparing herself for bed, she heard gunfire on the street outside the hotel and had pulled aside the curtains to see what it was all about, when the telephone rang. The voice over the receiver warned her to close the drapes or someone would shoot.

"Who could have been watching me?" she wondered, as she tossed and turned until dawn. The next morning, despite the Indian High Commissioner having told Kuku and those traveling to take only one overnight case, hundreds of Indians joined the convoy, toting large suitcases, shoulder bags, boxes tied with string and bags of goods.

"Can you imagine?" Kuku said, shaking her head. "So many things? I was told later that some Indians returned to Uganda to exploit the meager commodity situation to sell salt and sugar brought in from Kenya."

Kuku was classified as an official under the protection of the Indian Embassy, and rode alone in a vehicle with an Indian flag draped over the hood. "The trip was largely uneventful," she recalled, "although the convoy was halted two or three times at checkpoints and near the source of the Nile in Jinja. More and more apprehensive with each roadblock, the Indian High Commissioner blamed Kuku for the delays because of having kept "that boy." Fuming, Kuku retorted, "Then let me go back!" But the convoy continued through pouring rain to Kenya, where they spent the night before the motorcade moved on.

"When we arrived in Nairobi, Mama Miria greeted me with a display of much love and I was soon helping them with their household," Kuku recalled. Then, with a touch of melancholy, she continued. "Mzee was not with us, but I visited him at the residence of a retired Kenyan judge. He was very grateful. He told me that it had to be a foreign lady that rescued his son.

"Mama Maria wept like a child when I left Nairobi," Kuku remembered. "She told me we would write to one another and said God willing, we will meet again."

But Kuku never did return to Africa.

Chapter XVIII
Eradde:
"Is the Lake Calm?"

JUST AFTER THE COUP, the security situation in Uganda became worse. Despite the troubled times, often on a Saturday morning we drove with the children to the outskirts of town, to the summit of one of Kampala's scenic rolling hills. One such drive twisted near trees laden with *matoke* to a breathtaking view of Lake Victoria that stretched metallic gray as far as the eye could see. The children were drawn to the vista from Makindye and ran ahead of us to the old sundial under the shade of an ancient tree. We thought the tree had perhaps been planted at the turn of the century by an exceptional missionary couple, Dr. Albert and Lady Cook, who journeyed by foot on an eleven-week safari from Mombasa to Kampala in 1895 and made a name for themselves in the community at Mengo Hospital. So grateful were they for Lady Cook's medical assistance that the Baganda offered her a piece of land. But rather than select the productive lowland they suggested, she chose the crescent of a hill, which certainly would have been my choice as well, for black clouds of lake flies dipped and swirled over the water that harbored bilharzia-carrying snails. Malaria-bearing anopheles mosquitoes bred along the reed and papyrus shores of Lake Victoria and tsetse flies hid in the sodden swamps.

A sudden wind tossed the fishermen about in their wooden canoes on the lake far below. Soon large drops hurried us back to the car, just before heavy rain caused rivulets of water to course down the dirt road ahead of us. Careful to avoid drenched people walking on the sides of the road, covering their heads with banana leaves, Cole drove slowly toward the city roundabout with the clock tower.

“The clock’s still stopped,” the children chorused in unison, as we passed the grounds where Idi Amin had publicly executed his opponents.

Turning off Bombo Road, Cole had just suggested we stop for a cup of tea with Betty at Uganda Crafts, when, practically tripping in his haste to run down the stairs, Joseph, his driver, shouted, “Robin Biellik has been shot!”

“I need to go check out the situation,” Cole told me as Joseph backed up the Volvo to take Cole to Nsambya Hospital, where Robin had been taken to the operating theater and given a transfusion with blood from two Catholic nuns. Earlier, Cole had recognized the potential risk for his staff needing a transfusion and requested they be tested for HIV/Aids in order to ensure emergency blood supplies would be safe; however, the UN medical services in New York denied the request, saying HIV status was a personal matter and could not be known by the agency. Cole then conspired with Dr. Ray Gibbons, the UN doctor, who agreed with Cole that any UNICEF staff who came for normal treatment would confidentially have their blood tested for HIV.

In spite of having heard gunshots coming from the direction he was going, Robin, a nutritionist, started his car to drive to the market. Suddenly, a pickup screeched to a halt beside him. One of the thugs ordered Robin out of the car with his arms raised. Noticing the leather Swiss Army knife sheath Robin wore on his belt, he shouted, “Gun!” then shot him through the spleen. Robin staggered to the side of the road and watched helplessly as the men drove his car in the direction from which they had come. Realizing he needed immediate attention, he drove the hijacker’s pickup to the hospital, which was down the same road. The police, who assumed the thugs were still driving the pickup, fired at it, until they realized a foreigner was behind the wheel.

Although the surgery to extract the bullet went well, Cole put Robin on a plane for London. His intention then was to meet Robin’s flight at Heathrow Airport, but those in charge of security were of a different mind. Undeterred, Cole slapped his ticket on the counter and told the British Airways clerk he would like to board the next flight to the U.S. He then checked through security as though he would travel, and found the gate where an attendant was taking Robin out of the lounge in a wheelchair. Accompanying Robin through immigration and customs, Cole met with Robin's mother before he returned to the same clerk to change his ticket and get rebooked on his original flight to New York.

Thanks to excellent care, Robin recovered completely.

Across the Lines of Civil War

The six months President Tito Okello led his unruly government was thought to be the most turbulent period in Uganda's history—a time of intense insecurity, with the country in a relentless steep, steady decline.

Mike Woldridge, the BBC East African Correspondent Obote had banned from Uganda, asked Cole whether he could fly to Kasese with him on one of the Corridor of Peace flights. Cole agreed, and when Mike arrived from Nairobi, he took him directly to a waiting chartered flight. Mike interviewed Cole on the aircraft someplace around Mubende, just as they were crossing the front line. That night, Mike filed a story for the BBC from our home in Kampala, based on an interview with a child soldier they met at the landing strip in Kasese.

The next morning, the military police picked Cole up in a black Mercedes-Benz and took him to Tito Okello's nephew, Olara Otunnu, the Minister of Foreign Affairs (Cole had known Olara when he was the Ambassador to the United Nations under the Obote Government).

"Why are you supporting the rebels?" Olara asked Cole.

Cole replied that he wasn't supporting the rebels, but was providing medical kits to health centers within the country of Uganda.

"Who gave you permission to fly to Kasese?"

"Basilio Okello gave me permission under implied consent and I notified him about the flight, just as I notified the NRM in Nairobi," Cole replied.

"These flights must be stopped," Olara said angrily.

"Are you sure you want to go this route? Cole retorted. "If you stop us from flying within Uganda, we would then do cross-border operations from Rwanda or the Congo, over which your government will have no knowledge and certainly no authority whatsoever. This is something Ruhakana Rugunda (the right-hand man to Museveni) wants. Cross-border operations is the easy option for you," Cole continued. "If you allow flights to originate from Entebbe it implies you and your government trust UNICEF and that you are providing badly needed medicines to all the people of Uganda."

Olara reflected on what Cole said. Then he advised him to be very, very careful, as the military was upset with the interview Mike Woldridge had broadcasted over the BBC.

The background to all of the above commenced before the Okello coup. Confident and outspoken about children's rights, Cole was able to pull off the Corridor of Peace because he had created a following within the civil service and society, with aggressive programs in the area of health and immunization and rehabilitation at every health center and district hospital in the country. He also had a network of personnel with radios and vehicles across the country and was therefore well-informed.

President Obote was not immediately forthcoming with permission for UNICEF to work across the lines of civil war. However, after extensive discussions, he agreed, and advised Cole to get permission to fly from the commander of the military, Basilio Okello, before he flew again.

Shortly thereafter, the Okellos overthrew Obote's government, but since the flights had been initiated before the coup, Cole simply continued them. On one of these flights to Kasese, Cole met with Yoweri Museveni, and eventually the combination of the rumor mill and intelligence caught up with him. When he arrived back in Entebbe, the military police told him to report immediately to Basilio.

"You know, it was Obote who gave me permission to fly," Cole told the angry army commander. "He was aware of the flights, and as a politician, he asked me this question: 'What is the difference between a lady and a politician?'" But before Basilio could think about it, he told him, "When a lady says no, she means maybe. When a lady says maybe, she means yes. But when a lady says yes, she's not a lady. But when a politician says yes, he means maybe. When a politician says maybe, he means no. And when a politician says no, he is not a politician."

Basilio chuckled and, with the mood lightened, Cole left the commander's office reasonably confident that UNICEF would be able to continue the flights.

On an unsettled afternoon in November 1985, Araya Assefa radioed Cole from Kasese. "It appears they are all safe," Araya told Cole, referring to the five crew and forty-five passengers on the Fokker Friendship commercial flight hijacked in Kitgum District (Okello's territory) and flown to Kasese, the heart of NRM territory. Having driven to western Uganda via the West Nile, Araya had crossed the border into Congo and skirted the western side of the Ruwenzori Mountains before he crossed the border back into Uganda.

"Most likely the NRA was responsible for the hijacking, in order to stop the transport of military troops into the West Nile," Araya surmised when he met Cole in Kasese, where Cole frequently flew with the flights to assure that everything was going well as the military was convinced they were delivering arms to the NRM.

With mounting fury, once again Basilio Okello, who was now the Minister of Defense, summoned Cole to his office.

Expecting the worst, Cole called on Daudi Taliwako, the Chief of Protocol at Foreign Affairs and the UNDP Representative for some sort of insurance against being imprisoned or deported en route to Okello's office. Then, armed with nothing more than healthy fear and official documents such as the signed UNICEF Basic Agreement that dated back over two decades and the Ministry of Foreign affairs travel authorization to fly to Kasese, he reported to army headquarters. To add to Cole's concern, that morning the American ambassador had advised Cole to "leave the country with your family immediately."

"Okello was red-eyed," Cole told me later. "He was so displeased his voice shook. He was very clear with his instructions, and warned me never to fly into guerrilla-held territory again!"

Forced to his feet by his own convictions, Cole marshaled every ounce of self-assurance he had and reminded Okello that the flights were legal and within the context of UNICEF's mandate to help children. His own anger now dispelling his fear, Cole threatened to halt all UNICEF assistance within the entire country and to evacuate his international staff because of the country's rapidly declining security situation. Taken aback with his outburst, Cole waited uneasily—uncertain as to Okello's reaction. Fortunately, his words were an effective ploy. Aware that his government was in disarray, and that bilateral aid had been reduced and some United Nations technical assistance personnel had already been evacuated, Basilio suffered even more international disfavor than that of Obote, who he had so recently overthrown. On top of this, Basilio and Cole had a history of confrontation, for in 1981, as the Army Commander for Kampala, Basilio had been humiliated into returning three hijacked UNICEF vehicles. At that time, Cole had evacuated all international staff to Nairobi and shut down all programs countrywide; resuming them only after the vehicles were returned one week later.

Cole felt uneasy when he left Basilio's office, and decided to heed the American ambassador's advice to leave the country. The children eagerly packed their bags for the unexpected safari, while I left last-minute instructions with Pelesia, who reminded me she would repeat what I often told her: "Don't worry."

But worry I did.

Twenty minutes before our departure from Entebbe airport, Okello's nervous UNLA security forces feared another defection and shot down one of their own helicopters, which had been loaded with supplies for Masaka barracks. Earlier, a planeload of Okello's air force, which had taken off from Entebbe, had defected and landed in NRA-controlled territory with weapons and ammunition for the frontline UNLA troops outside of Masaka. As the tension escalated, so too did the frequency of Cole's visits to the offices of the Minister of Defense and the UNLA military officers, who were even more suspicious of the UNICEF mercy flights.

It had just rained, and we hurried our children over the wet airstrip to board the Cessna 402. Minutes later, Colin Stewart flew the plane low over the choppy waters of Lake Victoria in order to minimize the chance of being hit by the two antiaircraft guns located on two small hills either side of the runway. Holding my breath, I breathed a sigh of relief when the plane lifted into brilliant blue and the atmosphere lightened. Only then did I take notice of the passenger strapped in the seat next to me. Obviously distraught, the woman turned her pale face toward the window, but not before I saw her tears.

"What is it?" I asked, reaching out to hold her hand. Between sobs, she told me she had been attacked and raped. "This is why I'm being evacuated to Kenya, because I simply cannot cope with the insecurity of it all," she said, before she told me how she had barely survived her own attempt to end her life.

Soon after our arrival at the Hurlingham Hotel in Nairobi, Mike Woldridge held a press conference with Cole concerning UNICEF's flights to rebel-held NRA territory.

"In contrast to cross-border operations," Cole stressed in the interview, "the corridor-of-peace flights were flown with the consent of both Okello's government and the National Resistance Army." Emphasizing that he had seen large numbers of child soldiers with the NRA in Kasese, Cole reported many of the gun-toting *kidogos* were under the age of fifteen.

"The children are being trained to kill their own countrymen," Cole said. "They've been armed to fight an adult war."

Woldridge later flew to Uganda to meet with Cole, after which commentaries concerning the *kidogos* were featured on the BBC African news. Okello's government then made the most of Cole's objections to using child soldiers to criticize the NRA and took credit for paving the way for a further forty-one Corridor of Peace flights to carry medical supplies to rebel-held health centers across the lines of civil war.

Several restful days later, we shook our children awake and left Nairobi to drive back to Uganda through the magnificent "scar on the face of the earth so long and deep that it is visible 90,000 miles out in space," as Colin Willock writes in *Africa's Rift Valley*. Traveling to Kisumu over hills covered in tea plantations, we crossed the border into Uganda and, used to it by now, talked our way through roadblocks and checkpoints. We reached Kampala that afternoon, to an exuberant welcome from our animals. Barking and rolling over each other, the dogs sped down Mbuya Hill past Punda, who pricked up his ears. The cats meowed, Noey flicked her tail near the gate and Ibitz the duiker made us laugh by butting She-goat. Only the chickens ignored our delight at having returned home safely.

The situation in Uganda continued to plummet during the last two months of the Okello regime. The unruly UNLA forces were aware that Museveni's army was moving toward Kampala, and intensified their looting of vehicles and whatever else they could carry away from townships and villages. Wary of the security situation, which had become critical, and soon after a grenade was thrown into a crowded Wandageya bar killing thirty people, Cole decided to evacuate nonessential UNICEF staff to Kenya. Those who worked in upcountry Soroti had to be airlifted to Entebbe before they could be flown on a commercial flight to Nairobi.

"There's no trace of your guys," Mike Grennell, an experienced East African bush pilot flying for International Christian Aid radioed Cole, while flying over Soroti, northeast of Kampala. "They're not there. I don't see the Land Rover," he said, referring to the four UNICEF men, who had been threatened at gunpoint by a young man who asked his accomplices, "Shall I shoot them?"

Mindful that the Land Rover would be an easy target, Cole picked up the radio messages from the UNICEF men from our Mbuya home and simultaneously tuned in to the aircraft frequency. While Grennell circled the plane, Cole instructed Tony Espey to proceed to Soroti Airport for immediate evacuation. At first reluctant to leave his hiding place on the outskirts of town, as he was in charge of the large borehole drilling program that had stores of supplies, Tony finally agreed to drive on to the tarmac at the same time the plane touched down. Landing short, and well away from the heavily guarded hangers, Mike loaded his passengers and flew back to Entebbe.

That same evening, our family's conversation around the dinner table revolved around the men's evacuation.

"At least they're out of Soroti," I said to Cole as he scraped the last bit of tomato sauce from his plate with a crust of pizza. Gathering up the plates, he carried them to the kitchen to wash up, before he brought out the cake to celebrate Sanjeeb's birthday. It was our custom on Sunday evenings to play hide-and-go-seek, joined by our Dalmatian, who exposed our whereabouts by nipping and barking at our heels. That night, a perfect half moon tipped stars toward the children as they chased each other over and around the hill.

It was Cole's turn to hide. Unable to find him, the four of us called out, "Give up!" as "*grrr—rrup, grr—up*" sounded from our little swimming pool. Curious about the sound, Jessica walked near the edge of the pool. Suddenly, a wet arm emerged from the water and grabbed her leg. Shrieking, she lost her balance, and Cole leaped from the pool and lunged toward us. Breathless with laughter, I tripped over a root and was caught to be "it," but not until we bundled shivering Cole into a blanket, for he had been standing chin-deep in the water for ten minutes, observed only by the resident frogs.

Chapter XIX
Uganda's Wounded State

ON 20 JANUARY 1986, Cole met with the NRA leader, Yoweri Museveni, atop the highest hill in the region of Masaka. During their discussion, Museveni picked up his transistor radio and tuned in to the BBC. Listening attentively to Robin White's *Focus on Africa* report, which told of escalating fighting between his and Okello's troops, and that it seemed the UNLA was just days away from total defeat, Museveni turned off the radio and smiled. "I will soon make my final assault on Kampala," he said.

"What will you do with your child soldiers when the war is over?" Cole inquired, concerned with the prospects for the children used and trained by the NRA.

I too was familiar with child soldiers: just down the hill from us, the roadblock was often "manned" by young boys who fought for the UNLA. Far too young to be participants in battle, two scruffy kids stood stiffly with large rifles near the road I took into town. One, who wore pink trousers rolled up his bony legs, shifted the AK-47 he held and looked away, but not before his shy smile gripped my heart.

In *War, Violence and Children*, edited by Cole and Magne Raundalen, another child soldier was described as "solemn and wary." Peter was thirteen years old, but had the appearance of an undersized nine-year-old. He didn't smile. Neither did he talk. The guerrillas were his family. He knew no other. Considered "a good shooter" by his comrades, Peter remembered little of his childhood or family. He had never tilled the soil or herded his family's goats. His sole experience had been fighting, hiding, wandering, looting and surviving in the bush war.

John was portrayed in the same book as a confident, cheerful fifteen-year old boy, who joined the guerrillas after his parents, sister and brother were killed. "The guerrillas taught me pride for a cause, and what I know of discipline," he said. "I do not beg or steal." But his experience was not one of peace.

Children are not supposed to die, ". . . and they certainly are not supposed to kill," Alessandra Stanley wrote in a special report for *Time* magazine entitled "Child Warriors." Explaining that infantry evolved from the French word for child (it's amazing that it took until this century for international law to think about the use of children in armed conflicts), Stanley ardently reports that "children as young as eight fight enemies they do not know for causes they barely understand. For war does not rob a child of youth so much as it reveals his (her) innocence: ignorance of death and a nervy imperviousness to danger revealed in a boy's grin when a mortar shell falls close or in his eagerness to fire when instinct should tell him to duck. Children," Stanley concludes, "may not make perfect soldiers, but they make perfectly good ones."

There were many child soldiers with Museveni's troops as they continued their steady advance toward the capital, at a time when a knock on the door could mean a death warrant. A time when increasingly desperate UNLA troops were deserting to the north in stolen vehicles and all the loot they could carry.

Following a succession of car-jackings, on a January day in 1986, I turned my head in revulsion at the sight of four charred corpses grotesquely bloated on the pavement near the corner of the derelict Imperial Hotel. The bodies lay not far from the smoldering chassis of a twisted vehicle. Sick with the horror of it all, I parked the car on the side of the road and, although terribly aware of how vulnerable I was, sobbed until, drained, I drove slowly back to Mbuya.

Riveted to the disturbing scene below Mbuya, we watched desperate wives and children of UNLA soldiers flee like *siafu* ants, carrying their possessions and bundles on their heads toward whatever conveyance would take them away from the retaliation they feared. Surely the Mbuya barracks just below our home and farther up the hill would be completely deserted by the time the NRM troops arrived. Overhead, helicopters like giant insects with iridescent wings droned and hovered in the direction of Masaka. We waited for the NRM to take the city of seven hills and relieve us of the tyranny of the UNLA and the Okello's ragtag government. Soon after dusk, Kampala's residents bolted their doors to wait. All the while, we heard the faint sound of singing coming from the Verona Fathers' church just below our compound, with congregants asking for God's protection.

Museveni's Entry

Caught up in the fearfulness of impending war, we relied on radio messages to keep us alert to the unfolding situation. Aware that Museveni's troops had fought difficult battles at Lubiri and Makindye, we had been warned there would be heavy fighting in Kampala and when they reached Mbuya barracks. In preparation for battle, Okello's men had their tanks positioned at road junctions and the tops of hills. Thankful to be on the outskirts of the city, the next morning, from our position on Mbuya Hill, we watched the battle explode and smoke spiral toward the blue sky. Throughout that stressful day, dozens of terrified people found refuge with us on our compound, until they felt the situation secure enough at sunset to venture to their homes.

"Multiply the decibels we hear from the gunshots and you will really sympathize with the animals," I told the children, as Tiger whined in distress. Cradling his great head, the children whispered reassurances and found him a "safe place" under the table in the kitchen, where he spent the rest of the day. Penning up the donkey and goats, we left the little duiker to scamper about the premises with the chickens. The cats seemed unperturbed, but Bender, our Dalmatian, stayed restlessly near us. Even Noey, my bushbuck, pricked up her ears with each explosion, as she pulled the tender leaves from her favorite vines, which the old man had thrown over the back fence. Amazed that, despite the shooting, the *mzee* had brought Noey potato greens, I paid him his usual sum and he hurried off up the hill, most likely to where he lived at the base of a large tree he had hollowed out. Having had enough of the greens, Noey licked the salt from my arm and I stroked her russet coat. Her beautiful eyes were as still as deep waters. Some weeks before, we had made arrangements to transport her to Murchison Falls and her freedom, but this was never to be.

Pretending caution that dreadful morning, my beautiful doe trod after me with delicate steps. Suddenly, bullets sped past us and hit the brick wall of our guesthouse with a "*craaaccck*!" Noey bounded away. I watched our Canadian pediatrician friend Don Hillman (who with his wife worked with UNICEF and taught at Makerere University and Mulago Hospital) dive through the door of the guesthouse and under the bed. "Was that an attack on us or a stray bullet?" I called out from the kitchen.

Minutes passed before Don looked warily around and walked across the verandah to join me in the kitchen. "That was close," he said as we laughed uneasily together and decided to organize a ping-pong match with the children to ease the stressful situation.

That night, the children gathered sticks to make a fire under the flowering bombax tree. While artillery rumbled in the distance and bursts of machine-gun fire disturbed our senses, I laid an iron pan on the coals and cooked chapattis, which we ate with meat and vegetables. Watching the embers die, we took comfort in prayer and each other's presence, and before we tucked the children into their beds we hugged them close and began our own preparations to rest. But, although we were unbelievably weary, we found it impossible to sleep. The therapy of talk had long since ceased to calm my fears. Suddenly, a barrage of bullets ripped the air close by.

Terrified, I held my breath and reached for Cole. "Do you think we are being attacked?" I whispered, when a second volley of gunfire roused us from our bed.

"Come!" we shouted as one.

Awakened by the blasts, the children scrambled to the hallway, the only location in the house protected by four walls. Horribly conscious of how defenseless we were, we clung to one another as thuds of artillery continued, along with screams and shouts coming from the barracks beside us. Finally, after what seemed an interminable amount of time, there was a lull in the shooting and Cole crawled toward the window in the front room (which still held bullet holes from the last war) to peer through the glass. Still on his knees, he told us the night revealed nothing, just as another burst of gunfire sent him scurrying back to where the children and I were huddled together. Long into that troubled night we waited, not knowing for what, until we thought it safe to return to our beds.

At dawn, I left barefoot tracks on grass wet with dew and welcomed the sunrays that brightened the bougainvillea splashed over the dining pavilion. When the others awoke, we unfolded canvas chairs on the brow of the hill and raised our binoculars to watch a single helicopter hover in the smoky haze covering the valley. The helicopter dipped out of sight below the buildings in Nakasero, then reappeared above the city. Smoke rose from the Ministry of Finance as the helicopter circled, then dropped again to the field near Nile Mansions. We thought it might be going to evacuate Okello. Meanwhile, radio reports informed us that shells had fallen on the wards of Nsambya and Mengo Hospitals, where UNLA troops massacred patients in their beds.

Small fingers sought mine as I leaned against the verandah wall. Folding our youngest into my arms, I hugged her close and questioned the horrors we were witness to.

"I feel everything at once," Jhorna whispered.

Jessica exclaimed, "Mama come and see! Do you see that green speck on Tank Hill? It has to be Liz," she said, focusing the binoculars on Don Hillman's wife, who was wearing her favorite green dress. Liz had followed Cole's advice over the walkie-talkie, and was hurrying down the dirt road on the hill opposite, to a house where other UNICEF staff members were assembled. We all cheered, but Don clapped the loudest when we heard her cheerful "I've arrived!" over the handheld radio. Don's euphoria and the sight of Liz had invigorated us, and we devoured the pizzas I had made.

That afternoon, emotionally, physically and mentally drained, we slept, until darkness brought further uncertainties. In the morning, Cole faced the chaotic streets to bring Liz to Mbuya.

"It took all I had to drive past people running beside an Acholi soldier tied to the back of a car being dragged through the streets," Cole told us when he returned home. "The poor man was still alive," he added, shaking his head.

Sickened, I replied, "Then the tide has turned," worried that tribal revenge would dig Uganda's grave deeper.

The tide had indeed turned, and on 27 January, jubilant throngs of people accompanied Museveni's soldiers down the streets. Cheers broke out from the valley below as NRA soldiers surged toward the abandoned flats and the uniport barracks that had been occupied by the defeated UNLA who had fled toward Jinja—their only escape route. Just below our compound, Basilio Okello's wife and children had taken refuge with the Verona Fathers, and soon a concerned Father George walked up the hill.

"NRA troops have demanded entry to the church and the Vatican residence," he told Cole. "They are after Okello's wife."

Aware the situation was serious, Cole was accompanying the father down the hill, followed by our Dalmatian, when, with a sudden turn, Bender dashed back toward the dense undergrowth at the back of the house and barked relentlessly at half-a-dozen men huddled near the fence. Fearful as to how the armed men would react, I heard Cole request our remaining police guards (the others had stripped off their uniforms and fled) to lie down on the grass, with only the barrels of their guns visible. He then approached the desperate men and asked them to leave immediately. The terrified men pleaded for asylum. But, recognizing the danger of the situation, Cole pointed to the guards lying in the grass with their guns, and demanded they leave.

"That was difficult," Cole said with relief and pity, as we watched the defeated men crawl back over the fence and disappear into the bush.

Minutes later, a frantic Basilio Okello, who lived across the road from us, panted up the hill to request a UNICEF truck in which to flee with his family and possessions.

"Impossible," Cole said.

Repeating his request, the Okello's shoulders visibly slumped when Cole restated, "I'm sorry, but I cannot allow you to take a UNICEF truck." And again, with pity, this time we watched the commander hurry back down the hill.

The next day, Yoweri Museveni broadcasted his first address over Radio Uganda. Battle-weary victorious soldiers were shuffling down deserted streets when Cole and I slowly drove through what had been the worst scenes of combat. The ruins of war were evident everywhere. Corpses were strewn alongside the road near burnt vehicles at the top of Kololo Hill. It took several days for the bodies to be cleared from the city streets. Spent cartridges that glinted brass in the morning sun were being swept into piles by men with expressionless faces on the road near Lincoln School. Cole stopped the car to pick up two large casings where a mortar had been intermittently pumping shells at the NRA.

"Keep them," he said. "It will remind us of the brutal madness of war."

Apprehensive about what we would find at Uganda Crafts, as Bat Valley had been at the heart of crossfire, Cole drove off the main road and parked near the UNICEF building. It was eerily quiet, with not even the sound of bats that normally roosted in the eucalyptus trees. Broken glass and debris crunched under our feet as we walked past the damaged tables and thatched umbrellas that had been the coffee shop. Then I saw Betty walk toward us, leaning heavily on her stick. Weak with relief, we hugged each other close.

"Did any of the UNICEF police guards remain during the battle?" Cole inquired.

"They fled," Betty told us, as she slowly limped back toward the shop. "Of course they feared the war and fled. But by God's grace, this was not destroyed," she said, gesturing toward the displayed crafts. Shoving aside shattered glass with her metal stick, she smiled wearily and admitted, "I truly would not have left the craft shop, no matter what the fighting."

"He helped me," she added, motioning to Christopher Mugambe, who stood balanced like a crane on his good leg, holding a spear handle he used as a crutch. "He defended the shop against three groups of looters. Even some of them were armed."

"We did not know who would now become the government," Christopher said, shifting his weight self-consciously. "But I knew me and my family's livelihood depended on what I get from selling the shoes and belts I make in our shop. Men were coming to loot. I was wise enough to enter Uganda Crafts by breaking the blocks that covered that window," he said, pointing to where he had replaced the bricks after crawling through the hole. "I took two Karamojong spears, and when the men arrived, I told them, 'This is a disabled place.' But they harassed me and I threatened them . . . and so it went until they left the place."

"For me, I spent many hours wondering when I would die," Betty said, shaking her head. "Every time I heard a cannon, I had to sit on the bucket and let loose my fears until there was nothing to pass," she said, managing to chuckle. "See that staircase?" She pointed with her cane to the side of the building Uganda Crafts occupied. "I watched from my hiding place as NRA soldiers ran up the steps to go the roof of our building. From there they aimed their guns on Okello's troops, who were shooting from the road opposite us. It was really too, too much . . . too very terrible," she said, collapsing onto the nearest chair. Sighing deeply, she concluded, "But let us pray to God that our problem is now over."

Through the years, Betty forged on to educate twenty-six children, many of whom were orphaned during the war. First leasing, then purchasing 110 acres of land on Lake Victoria, she started a village as a form of security with the profits she made. She also continued to run Uganda Crafts as an uplifting place—one that provided income for hundreds of women and suppliers.

In later years, an empowered Betty was elected as an LC5 Councilor for Mpigi District to represent the physically challenged. When I asked her what she hoped to achieve with her new title, she answered without hesitation, "I want them to unite with confidence so their voice can be heard—to fight poverty and ignorance and to be recognized by the public. I also would like to see policies that the government will put into place that will be friendly for the disabled."

Grave thoughts were temporarily pushed aside as Cole and I swung up Mbuya Hill. The morning was aquiver with birdsong. Glad to have us back home, our three children ran to the car to greet us and I blinked back tears with thanks to God for His protection.

A day or two later, the children were eager to return to school to see their friends, but returned that day to tell us only 40 out of the 160 students who had attended Lincoln before the Okello coup had returned.

It took time for peace to be restored. In their hasty retreat, Okello's forces hijacked vehicles, looted houses, set fire to villages and killed civilians. Concerned about the workshop and fleet of vehicles stored upcountry in Soroti, UNICEF staff welded shut the containers and hoisted the Scania trucks on hydraulic jacks after they had removed the wheels. Dropping some vehicles on their axles, they parked others with the drilling rigs and compressors in a makeshift corral they thought would make it more difficult to loot.

"You will be shot if you land," the men were told as the pilot and Cole circled over Soroti in a light single-engine Cessna aircraft to evacuate UNICEF staff. Despite the warning, the pilot informed the tower they were a United Nations flight and shuddered to a stop in the middle of the runway, near a Land Rover bristling with military personnel and rocket-propelled grenades. Disembarking with their arms above their heads, the men heard a military man shout, "You are under arrest!"

Disregarding the command, Cole, who had earlier communicated with Tony Espey over a walkie-talkie, jumped into the UNICEF Land Rover Tony had driven up at the same time the plane landed. Hurriedly driving to the UNICEF office, they collected cash and documents before they stopped to see the army commander to tell him their intention was to fly directly back to Entebbe. Back at the airport, the guards had their rifles raised, with orders not to let the *wazungu* go. Throwing caution to the wind, Cole told them he had just spoken to their commander, and the pilot lifted the plane into a cloudless sky back to Entebbe.

Once again, Cole had arrived home safely.

"I have come," Museveni told the public in his first radio broadcast: "Not only to change guards but to change a system that has caused damage and suffering."

When the euphoria of Kampala's residents abated a little, Cole met with Museveni near Republic House under a large old tree, thanks to the arrangements made by Araya, who had literally followed (at a safe distance) the NRA into Kampala. Because Cole had met Museveni in the bush and had had direct communication with him through a handheld walkie-talkie (courtesy of Araya and his friends in the NRA), he was the first person from the UN or diplomatic community to meet the new President. When Museveni told Cole he planned to meet the diplomatic corps at the President's office, Cole advised him not to be seen in the same position as Okello, but to meet the diplomatic community to distinguish himself with his own identity. The following day, those invited were seated two-deep in a semicircle under the tree, listening to the new President. Museveni requested that those present introduce themselves. When it was Cole's turn, Museveni pointed a finger at him and said, "For you, we all know you well enough."

Some days later, Cole chartered a Red Cross flight to Gulu to meet with the defeated Okellos at the Acholi Inn. Concerned about his family still within the walls of the Verona Fathers, Okello asked Cole to check on them. He did so, and eventually arranged transport for them to travel north when things settled down. However, it was not possible for Cole to recover the UNICEF vehicles Okello's troops had looted, for Museveni's troops were still pressing north and would soon drive the UNLA from Lira and Gulu across the border into southern Sudan.

Alice Lakwena

The appalling situation in northern Uganda was further exacerbated when the UNLA forces fled north to their homes. The returning soldiers did not easily readjust to traditions and the village way of life and added to the area's insecurity and disintegration. It was during this time that, at the age of twenty-nine, a woman by the name of Alice decided to wage a war against the evil her people had suffered. Raised an Anglican, Alice was selling fish in the town of Opit, thirty kilometers from Gulu, when she claimed that the Christian spirit of an Italian captain, Lakwena, took possession of her. Announcing that she had received the spirit after a light hundreds of meters high descended on her, she claimed, "People saw it. That is why they know I have the spirit."

In order to "purify" the 150 soldiers Alice recruited from the Uganda People's Democratic Army (UPDA), she had them sprinkled and immersed in water before, among other rituals, they were required to burn charms, spit into the mouth of a pig, which was later killed, and mold clay figures they chopped up and burned. She had her initiates memorize twenty safety precautions, which she convinced them would protect them in battle, and which, if not adhered to, would cause their death and defeat. The moral renewal she insisted her Holy Spirit Mobile Forces (HSMF) practice offered her soldiers a certain security. Alice reinforced their confidence by saying the spirits would decide who should die, and the stones she blessed would turn into grenades.

Although many of her men were killed, her troops, who depended on the local populace for medicine, money and food, managed to defeat Museveni's better-equipped NRA and marched from Kitgum to Lira, Soroti, Kumi, Mbale, Tororo and as far as Jinja, until, in 1986, the Ugandan army crushed Lakwena's followers and Alice fled to Kenya. She was arrested in Kenya, having entered the country without the required identification papers. In time she was released and was granted asylum with the UN High Commission for Refugees (UNHCR), and was later "seen in a small bar (in Kenya), wearing a white blouse and a blue skirt—drinking Pepsi Cola with gin," as Heike Behrend wrote in *Alice Lakwena and the Holy Spirits.*

The controversial Alice Lakwena died at age fifty-one in a refugee camp in northern Kenya, Kenyan newspapers reported in January 2007. At that time, President Museveni responded to Alice's father's request that her body be taken back to Uganda, and announced that his government would assume responsibility for Lakwena's burial in Gulu District.

Alice had inspired her cousin, Joseph Kony, who began his own armed struggle against the government. His force, the Lord's Resistance Army (LRA), destabilized, killed, maimed and dehumanized Ugandan citizens, causing living hell for the people in the region. "Spirit-led" Kony used fear to enforce discipline, and had his soldiers undergo initiation before they battled the government and other resistance movements. His troops were brutal, and used violent means to capture children's minds and bodies to fight and act as porters, servants and wives. "Thirty four people were burnt alive and another 18 were bludgeoned to death by rebels" when they attacked a village in Ogago, reported the *Daily Nation*. Shortly after the rebels left, an army intelligence officer arrived at the scene and took a picture that verified the dreadful story. Another village suffered an act of retribution because a boy who had been abducted from there escaped from the LRA with a gun.

Continued horrific attacks displaced entire populations in northern Uganda. Hundreds of thousands of people fled to towns and trading centers, which turned into internally displaced people's camps. Over thirty thousand children and young people—most between the ages of seven and seventeen—were abducted by the LRA over the years. Tens of thousands of other children trekked miles from their homes and villages to sleep in the safety of nearby towns. The local population suffered terribly with reports that the NRA was also known to plunder and rape. Meanwhile, Museveni used the war in the north to restructure and expand his army and to increase the country's defense budget.

"So many children from the LRA do not know what education is, but know how to fire a gun—they do not value life," my friend from Pabbo village in Gulu District told me one day as she poured orange Fanta slowly into a glass. We were in a restaurant located on a back street in Kampala. Lowering her voice, she bent her head closer to mine. "The LRA harasses people, kills domestic animals and even takes our children as wives—trains them for soldiers. The government's response has been to send the Uganda People's Defense Forces (UPDF) to the Trading Center where the people live near the forces in 'protected villages.' Now the people not only fear the LRA, but are frightened to report incidents that the soldiers are responsible for—information that might be picked up by the local radio station."

Ebola

The radio station Rose referred to was set up by Lacambel Wodogena and his colleagues to alert civilians in the Gulu area about LRA attacks and other security matters. The reporters later turned their listeners' attention to one of the most virulent diseases known to humanity (identified in the Congo in 1976 as Ebola), and addressed the frightened public's queries: "Where was the last death?" "Is it safe to share a cigarette?" "Will a handshake spread the disease?"

Esther Awete had done nothing out of the ordinary—she had just traveled to another village to purchase cassava leaves that she brewed into beer for small sums of money. Soon after her journey, she was gripped by a fatal fever, and died an agonizing death. Her sorrowful family bathed Esther's body in the traditional way and covered her casket with red earth. Tragically, a few weeks later, seven of those who had taken part in Esther's funeral died of the same deadly hemorrhagic disease.

Dr. Mathew Lukwiya was aware that he was dealing with a highly contagious illness and opened up an isolation ward for those affected at St. Mary's Hospital in northern Uganda. He then sent his patients' blood samples for analysis to South Africa. They came back with the result that the epidemic was indeed Ebola. Alerting the Ministry of Health concerning the confirmation of the deadly virus, the Acholi doctor warned his frightened staff to wear protective clothing while caring for their infected patients.

The LRA rebels had invaded St. Mary's Hospital earlier, at which time Dr. Mathew persuaded them to take him captive rather than the Italian nuns they had come for. Then, according to Blaine Harden, who wrote of the incident in the *New York Times*, the doctor "marched around the bush for a week in his gown before the rebels let him go," and later opened the walled compound at St. Mary's as a sanctuary from the rebels. About nine thousand people took shelter there, until Ebola scared them away. Admirably dedicated and courageous, Dr. Mathew continued to lead his hospital colleagues against the virulent Ebola, until he too succumbed to the dreaded disease. Ugandans grieved with his demise, for Dr. Mathew's words and actions demonstrated extraordinary care and serious commitment.

Chapter XX
The Hurt of Leaving

A STEADY BREEZE rippled the waters of Lake Victoria when Kenya Airways touched down in Entebbe. The sun's first rays warmed our backs and infused the lush green landscape with golden light as we walked across the tarmac toward the airport. Joseph was there to greet us and inquired, "How did you like the Sudan?"

Unusually subdued, he loaded our suitcases into the boot of the Volvo, aware we were back in Uganda only to pack our possessions.

Three weeks earlier, Cole and I had made a difficult decision to enroll Jessica and Sanjeeb in the international boarding school I had attended in South India.

The day we left our two fourteen-year-olds high in the western ghats, mist swirled through the eucalyptus trees. Sanjeeb circled with friends he had already made, but Jessica held her curly head down as she wiped away her tears. My heart ached for the children, who would now be responsible for their own decisions. Swallowing her hurt, Jhorna put on a brave face as she said goodbye to her brother and sister, with plans already in place to follow their footsteps to the highlands of Tamil Nadu. The school she would be attending in Khartoum did not offer the upper grades her siblings required.

Later, as we packed our belongings, Jhorna and I spoke about the emptiness we felt without her brother and sister. Stacking baskets toward the bottom of crates, we tucked rugs and fabrics around carvings, and sorted artifacts, brass and books into boxes, along with bedding, ceramics, clothing and kitchenware. Our pottery and furniture was distributed to friends. Our house now stripped of furniture, we ate our last meals on the verandah, where the flowers and the view of the lake still remained.

Our chicken flock had diminished to one splendid, arrogant

rooster, his hen and small brood that Pelesia promised to keep out of her cooking pot. Punda, Jessica's stubborn donkey, would soon pasture with Isaac's cattle. The goats would go to our gardener, Emmanuel, who now had a driver's license and would drive the Isuzu for Uganda Crafts. Bender wiggled nervously in the front seat as we drove to Sanjeeb's Greek friend's house over the hill. Lowering an eyelid over her pink eye, her blue one looked at me for a long minute before she bounded out of the car and dashed about the premise of her new masters. Clearing the lump in my throat, tears stung my eyes and I quickly drove away. But my real agony was still to come.

With grief too much to bear, I sobbed my heart out as Noey gently licked my face. Begging her forgiveness, I sought comfort in her beautiful eyes and held her close one last time before I penned her up with her favorite sweet potato greens. My tears continued to fall when our little duiker nudged my legs with her wet nose and I stroked her coarse coat. Tiger only groaned a farewell when I took his tousled copper head in both my hands. Looking into his tired eyes, opaque with cataracts, I thought my heart would surely break. Then, with sorrow I knew not how to share, I drove Jhorna to the Sunday evening prayer service, while her papa and the veterinarian painlessly put our animals permanently to sleep.

In church, the organist played the introduction to a familiar chorus, but I simply could not sing.

Our last night in Kampala, Lake Victoria shimmered in the distance from what had been our home. That night we shared *matoke,* steamed bananas, and stew with a few friends by candlelight as ". . . the lamps were packed away," Mort Rosenblum wrote in his book, *Squandering Eden, Africa at the Edge.* "After five years of internationally recognized work," Rosenblum continued, "Dodge did not say it, but friends said UNICEF had nudged him along to a promotion elsewhere. For he had made waves about the *kidogos,* the 3000 to 7000 soldiers still in the army.

"Talk was optimistic. One guest was Betty Kinene, thirty-four and crippled from polio since childhood. She had lost her husband in the

Luwero Triangle and been left with six children. Cole's wife, Marilyn, had helped her set up a shop in the UNICEF building to sell crafts made by handicapped women. The business thrived and Betty was a star. She was looking hard for funds to build a training center for handicapped women near Kampala . . ."

Wrapped in the warmth of that beautiful night, it was with a deep sense of sadness that I thought about leaving our friends and the situation we had become so involved with.

Chapter XXI
The Sudan

THE RIVER Geoffrey Moorhouse describes in *The Nile* as traversing such a variety of landscapes, such a medley of cultures and such a spectrum of peoples, flowed serpent-like through the desert far below the plane. It was easy to see the confluence of the White and Blue Niles as the pilot circled before landing in Khartoum.

Abdul Manjit greeted me at the airport in the gracious manner Sudanese are known for, and informed me Cole was due to arrive from London later that evening for our introductory visit to the country.

"The Sudan will be a challenge for you," the Executive Director of UNICEF had earlier cautioned Cole when we met with him in Nairobi in transit from Mozambique to New York. "It's the largest program in the Middle East and North African region, and up until now has not really had a UNICEF profile. Keep in mind that it will be a difficult country to work in."

The competition for the Representative's job had been intense, partly because it was a director-level position (The UN system has four categories of staff: GS—General Service, P—Professional 1 to 5, D—Director 1 and 2, and Assistant Secretary to Secretary General. Cole's job in Uganda had been a P-5), and partly because there were two other strong candidates vying for the post: Stephan De Mistura, a charismatic Italian-Swede who was working for World Food Program in Khartoum, and Jane Campbell, the head of the UNICEF Human Resources Division in New York, who had applied for the job because her British husband had just been appointed the High Commissioner for the Sudan.

The plaintive call of the muezzin summoned people to prayer as Abdul drove me down deserted streets to the Meridien Hotel. Dropping my overnight bag in the room Cole and I had been assigned to, I pushed open the window to let in the evening air. Dry wind tangled the lace curtains as I leaned out to watch white-robed men greet each other with hearty embraces. Across the road, behind low compound walls, people were stretched out on *angareb* cots in the oppressive heat.

Suddenly overcome with memories of leaving the children and Uganda, and the thought of living in this parched land, I threw myself across the bed and sobbed my heart out until I heard Cole's knock on the door.

In Emil Ludwig's *The Nile*, I learned that the Mahdi and his dervishes had stormed the city of Khartoum "with a death dance of delirious humanity before they took General Gordon's head to the victorious Mahdi as a trophy to set up in front of his house on a spear for the people to pelt with stones."

But then what was the Mahdi to do with the conquered Sudan? "In public," Ludwig wrote, "the Mahdi wore his linen robe, but inside his house, he wore costly stuffs and when he reclined on brocade cushions, a few of his wives stood behind him with ostrich-feather fans, others massaged him and the nearer to him they stood, the more costly were their garments. The water he washed in was sold to the faithful as well as little sacks containing the earth his foot had trod."

Following the Mahdi's death, his successor, Khalifa Abdullahi, made Omdurman his capital until Kitchener arrived.

I closed the book and watched dusty spirals lift and deposit litter around the deserted street. The hotel occupants opposite had their shabby curtains closed against the glare and relentless midday heat. Throwing aside the cushion, I sagged back in the cane chair and noted that the moisture had already evaporated from the shirts I had hung dripping from hangers. The gnarled branches of a bougainvillea plant in a clay pot beside me showed signs of sprouting. Finding a dipper in the bathroom, I watered the struggling plant as Jhorna emerged from the darkness of the hotel room.

"It will soon be too late for lunch," she reminded me.

We made our way across the sun-baked intersection to the dining room of the Greek-owned Acropole Hotel.

Slowly climbing the stairs to the second floor, we pulled our chairs near a table spread with a clean, checkered cloth and welcomed the whirring fans and air coolers that blew moist air around the empty dining room. It was near closing time and the old waiter robed in a light-blue *jelabiyya* took his time to shuffle toward us in loose slippers. The tribal scars cut down his handsome face deepened when he smiled and inquired, "*Intum kwaysiin*?"

"*Alhamdulillah*," we replied in unison, acknowledging the Sudanese reverence for God.

A little over a year later, the Acropole dining room was crowded when, just after 7:30 p.m., a man pushed a briefcase filled with explosives toward the table nearest the stairs and fled. Chris and Clare Rolfe and their two young children, who were passing through Khartoum on their way to work with a small British charity in Port Sudan, were seated at the table. They were blown to bits.

The same fateful night, Rose, our friend from southern Sudan, was enjoying dinner with friends when she heard a *hisssssssssssss*. She was hurled through the damaged roof to the street two floors below.

Palestinians who claimed to be members of Syria's Abu Nidal planned a simultaneous attack on the Acropole and nearby British Club "on the anniversary of the day Israeli immigrants began to settle in East Jerusalem," the *Sudan Times* reported. But the explosion at the hotel alerted those at the Sudan Club, who hid in their rooms before the terrorists stormed the club. Seven people lost their lives at the Acropole, and many more of the twenty-five injured Americans, British, Dutch, Canadians, Swiss, Poles, Bangladeshis and Sudanese would have been killed if the ceiling of the hotel's dining room had been concrete rather than corrugated iron sheets. Dazed, deafened and confused, the survivors stumbled from the building.

Rose regained consciousness in the hospital, where good Samaritans had taken her. She was slow to recover her mental equilibrium, but had miraculously suffered few physical scars. Equally fortunate, Charles, a UNICEF employee who was thrown to the pavement while walking on the sidewalk near the hotel at the time of the blast, suffered only bruises and recurring fear.

Two days after the attack, Cole and I drove to the Acropole to see the damage for ourselves. Avoiding the gaping hole where the Rolfe family had eaten their last supper, we greeted the old men somberly sweeping rubble into piles. Overhead, the sky showed blue through the irregular outline of what had been the ceiling. Unscathed on the far wall, the framed face of Christ looked sorrowfully on the destruction.

Simmering Tensions in Sudan

Prior to being elected as Prime Minister of the Sudan in April 1986, Sadiq al-Mahdi had been imprisoned by the former President Nimeiri for protesting the Islamic code of laws that included hand amputation for theft, flogging for alcohol consumption and stoning to death for adultery. Aware of the growing public's discontent and his government's survival, al-Mahdi declared he would allow the Muslim majority to choose the laws that governed Muslims, but that the rights of non-Muslims should not be violated. Many southern Sudanese adamantly opposed *sharia*, the fundamentalist Muslim laws.

An estimated 100 percent annual inflation rate sent prices skyrocketing and created severe commodity shortages. Disgruntled citizens queued outside Khartoum's bakeries for *aeesh*, bread, from before dawn. Dealing with frequent electrical shortages and fearing theft from the less privileged city residents, our neighbor suburban dwellers padlocked their rattling generators into metal cages next to pools of black oil.

One morning, hearing a commotion from the street below, I pushed open the kitchen window of the flat we had rented in Khartoum II to watch protesters burn tires to express their grievance against the scarcity of provisions. Waving banners, they shouted, "*Sukkar maa fii, aeesh maa fii*"—"No sugar, no bread"—as they marched past goats with cotton bags tied about their udders, who were nibbling at scraps of paper strewn on the side of the road.

Fatuma joined me at the window to watch the demonstrators go by. Mumbling Arabic words I could not decipher, she turned back to the kitchen table to stir sugar into a large pitcher of *aradeib* she had made from the seed of the tamarind tree. Placing the drink in the refrigerator, alongside jugs of *karkadeh* (dried petals of roselle flowers) and *gangaliiz* made from the seed of the extraordinary baobab tree, I thanked Fatuma. Pulling her cotton wrap over her head, she headed to the bathroom to perform ablutions, then spread a small cloth on the floor on which to say her prayers. Then, with slow deliberate steps, she returned to eat the *fatuur*, breakfast I had prepared for her: eggs, *aeesh,* fresh cucumbers, onions and tomatoes. Washing it down with very sweet black *shay*, she smiled and said, "*Shukran*."

Saleh had introduced Fatuma to us. When I agreed to hire her, he had lifted both his hands toward Allah and repeated, "*Shukran*" many times over. Assuring me he knew her well, Saleh, whose work as a *ghaffiir* was to guard UNICEF's premises, lowered his voice and told me Fatuma had been forced to leave her earthen house in a small village in Kordofan "because of no rain."

The following day, someone pressed the doorbell over and over. I went to the door to admonish the caller, but Fatuma's merry eyes met mine and she followed me into the room. Removing her pink *tobe*, she straightened her faded loose shift and with gestures inquired where she could place the small bag she carried. Then, slipping out of her worn rubber slippers, repaired with plastic wire, she smoothed back her gray hair, which was twisted into thin wispy braids, and gave me a wide smile before she took my hands in hers and said, "*Allah karim*," God is great, again and again and again, with the same tempo she had rung the doorbell.

Not at all sure what was expected of me (as the woman I greeted was not at all like the ones who had worked for us in the past), I offered Fatuma a cup of tea before I presented her with a mop. She pushed it around the floor with her foot, leaving rings of dirty water and the stick to trail on the floor. Confident she would soon learn, I showed her how to wield the mop and told her which rooms I wanted cleaned, and left her to her task. Much later, she found me washing up dishes in the kitchen and pulled my arm to show her where to collect her bag. Waving goodbye, she proceeded up the steps to our bedrooms.

Summoning her back down the stairs, I accompanied her to the front door, where she left me with, "*Bukra, insha'allah,*"—tomorrow if God wills.

Fatuma and I rehearsed this exercise day after day, until I lost the bet with myself that she would learn our alien ways. I longed to communicate with the old woman. It was obvious she managed with very little: the clothing she wore, perhaps a mat and a few pots and pans.

The large flat we rented in Amarat was not what we had hoped to locate on the banks of the Nile, but it was conveniently close to the international school Jhorna attended. Just as we were getting used to living without furniture, our personal effects arrived from Kampala, box by box, with articles missing or broken. *Nus Nus* and *Babanus*, the two Siamese cats we acquired, soon took over the *angareb* cots we placed on the flat roof next to pots of plants. I watered the plants copiously, determined they would grow despite Khartoum's blistering heat, wind and dust storms.

Fatuma was delighted with the jewelry I had collected from various countries through the years. She draped the necklaces made from shells, glass, silver, copper, assorted seeds and cowrie shells around her neck, before handing them to me one by one to hang on a rack on the wall. When we were finished, she rocked to and fro on her cracked heels and gazed for a long while at the picture I had hung of an African woman balancing a calabash of grain on her head. Tracing the woman's face with her weathered hand, she looked at me and smiled.

In time, with less energy to supervise and feed Fatuma as I had become involved with women's work, I had a friend translate my decision to relieve her of her services. Concerned about the old woman's reaction, I was reassured when Fatuma gave me a cheerful grin. Patting my arm, she reminded me, "*Allah karim,*" before walking slowly and with dignity toward the cupboard where she kept her few possessions. Gathering the items I had given her, she left the house without a backward glance, perhaps with praise to Allah that she did not have any more *khawaja* floors to wash. Within a few days, a young Ethiopian cousin of a friend of ours took over the housework in a more competent manner.

Al Sudaniya

Across the river from Khartoum, in the heart of Omdurman, the Mahdi's tomb rose above the surrounding low dwellings at the side of the dusty road. Parking the Land Rover near the souk, Cole and I walked past *dura* (sorghum), dried herbs, tamarind seeds, beans and lentils stored in jute bags for sale in small shops. The merchants greeted us with, "*Salaam aleikum*" and smiled when we responded with, "*Aleikum as-salaam*," accustomed by now to the lengthy Sudanese exchange of greetings.

Donkeys plodded slowly down the dirt road, pulling carts laden with vegetables to be sold or delivered to market stalls heaped with vine-ripened tomatoes, purple eggplants and bright yellow limes. Farther down the lane, an elderly man with a henna-streaked beard sold beads of all varieties and cheap ornaments brought from distant lands. Selecting a strand of bronze-colored amber strung on leather, I paid the trader and hung the necklace about my neck. Disappointed not to have found more ethnic items in the market, I crossed the road to a *dukaan,* where goat hides strung with cowrie shells hung from the wooden beams. It was cool in the dark shop. When I inquired whether he had any local crafts, the merchant shook the dust from beautifully woven baskets he had stored overhead, and told me they had been made by women in the western part of the country.

That afternoon, I worked on a proposal with objectives that would encourage creative producers, particularly women, to be paid fair wages as soon as goods were received. Designed to encourage the rotation of profits into funds in order to establish branches in other regions of the Sudan to help rural women with immediate cash, my proposal was accepted by Alastair Scott-Villiars. Through Band Aid, he agreed to fund the market outlet with a $250,000 grant for craftswomen that four other committee members and I would administer.

While We Breathe, We Hope

Following the opening of the craft outlet, enthusiastic suppliers brought blankets, baskets, jewelry, clothing, furniture, instruments and numerous other items to the old house shaded by large trees that we rented on the corner of one of Khartoum's quiet streets. Soon, Al Sudaniya attracted eager customers, who also enjoyed traditional drinks and *gahwa*, coffee, served under thatched umbrellas made by Shilluk men.

It was a stifling afternoon. The wind blew dirt into my eyes, and played havoc with Fawzia and Shahwa's *tobes*, which they held over their faces. Not a tree or blade of grass was in sight.

Dust clouds deepened the color of the sun as Mohammed, Al Sudaniya's driver, rattled the pickup over the rutted road. Slowing the vehicle through a deep ditch, he drove around a labyrinth of low mud houses known as Haj Youssef, home to thousands of refugees who had fled the war-torn south of Sudan. Parking near a newly constructed building, Mohammed showed the way to a small, dark room where women were gathered to discuss how they could market what they made.

Fine gray dust filtered through the cracks of the shuttered windows as we strained our eyes in the dim light to look at the articles the women had sewn. On the far side of the room, Rose tucked her long legs under a low stool and asked whether we would pay cash to her struggling sisters. Then, turning to me, she said, "While we breathe, we hope." For it was not a coincidence that the displaced southerners had not received their share of the country's benefits.

From that day on, "me Rose," as she referred to herself, was my friend. If it was too late in the day for her to take a public conveyance, we would drive her back to the little room she shared with other southerners across the bridge, past Omdurman. Or, if she required a bed for the night, she slept on the *angareb* in our front room.

Determined the displaced would have access to sell what they had made at Al Sudaniya, Peter, from Oxfam, and I followed Mathew's stilt

-like gait to the rubbish heap on the outskirts of Khartoum, where about seven thousand of the two million southerners forced to seek refuge in towns away from the south were allowed to subsist. Picking our way around refuse and structures made from discarded cartons and whatever sacking and material they could find, we paused for a moment to speak with those curious as to who we were.

We had to stoop to enter one of the larger huts, which Mathew's brother had pieced together with cardboard and plastic over a metal frame. Gabriel welcomed us, then introduced us to the once-proud cattle herders who had roamed the plains and wide spaces of the Bentiu area and Upper Nile, who were now reduced to long useless hours or working day-labor if they could find it. Folding their long limbs in order for us to crowd together with them on the dirt floor, the Dinka men soberly nodded their heads in greeting and told us their meager earnings were brought in by their wives who brewed *merissa*, local beer, from *dura*.

"We eat millet when we can get it," Gabriel said, drawing deeply from his clay pipe, the dark outline of his face silhouetted against the sunlight that shafted through the tattered piece of plastic overhead. "Our water is purchased from a man who brings it to us in on a donkey cart," he added, blowing smoke into the room.

Attentive to the expressions on the men's faces, I heard Peter stress the necessity for them to retain their self-respect and traditions.

"Our idea is to provide you with loans or the equivalent of six hundred pounds to purchase small barrels and skins to make drums to sell through the craft outlet this lady is setting up," he said, motioning to me.

"For sure," I added, determined to give these men some sort of hope, "the profits will be yours at Al Sudaniya."

The men then talked softly among themselves, then shook our hands and asked for time to discuss the matter, which they felt held promise. For all of us knew it was possible for branches to sprout from cut trees.

* * *

Heat hung heavily on our energy and Al Sudaniya's activities one afternoon when a woman dressed in Ethiopia's traditional white dress slipped quietly through the rear entrance of the shop. Careless with what she selected, she filled the large basket she was carrying, then bent over to say something to the child clutching her skirts and left the premises.

Huda left her position at the front desk and hurried after her, calling out, "But you have not paid!"

Anticipating trouble, Saleh, who had transferred his *ghaffir* loyalties from UNICEF to Al Sudaniya, hurried across the road to bring help from the police. Meanwhile, I guided the distraught woman back into the shop with the intention of reasoning with her.

"I want peace," she kept repeating through tears, as customers gathered around. "It is only peace." Flinging her arms to include the merchandise displayed, she threw passionate words at me. "All of this belongs to all of us—we own everything. Aren't we of the tradition that made it?"

Glancing sideways, I caught sight of Saleh escorting two officers into the room, and advised the woman to quickly leave the shop.

Slow to heed my warning, the woman resisted the policemen who forced her to stand. Spitting and clawing, she cursed the officer, who pulled her kicking and screaming out the door, followed by the terrified child. Examining the handbag she had left behind, I found the woman's work permit and other documents that indicated she was a refugee in the country. Hurrying toward the Al Sudaniya vehicle to follow the police, I looked back to see Dr. Hamdi hobbling after me, leaning heavily on his cane.

"*Maaleesh, maaleesh*, Marrrilyn," he said, rolling the r's of my name. "Never mind, please."

Frustrated and angry, I told my old friend that I must go after them—for what would happen to the woman and the child?

Impatient now, Hamdi took my arm and firmly led me back to the shop. "You must be careful not to take part in such matters," he told me in a paternal way. "Remember, you are a foreigner here."

I remembered his words when, not long after, security men questioned two of Al Sudaniya's employees as to the activities of the shop: Did we entertain political parties? Did refugees—Ethiopians, southerners, Dinka from Hillat Shook bring goods to sell? Were they allowed to enter the premises freely?

Victims

Considered rebel sympathizers, the Dinka from the south were the victims of northern discrimination and an expanding bitter civil war subject to militia harassment, slavery and famine. With resistance from the north, John Garang recruited men for the Sudan People's Liberation Army, primarily from his Dinka tribe—furthering the hostility and suspicion felt toward the cattle herders. The background of the conflict went back to 1955, when the Anyanya, led by the Southern Sudan Liberation Front, waged a bloody "bush war that lasted for 17 years until Khartoum ceded the region limited autonomy under terms of the Addis Ababa agreement in 1972." When oil was discovered in the Bentiu area in 1981, it became clear that oil exploitation would bring few benefits to the region and SPLA rebel activity increased both "at Bentiu and on the massive Jonglei Canal, a project designed to deliver a greater volume of southern Nile water to irrigation projects in the north and in Egypt."

Announcing the re-division of the south into three self-governing regions, thus nullifying the Addis Ababa agreement, the last pillar upon which the uneasy southern truce still balanced, and imposing sharia law, the then President of the Sudan, Nimeiri, confirmed southern suspicions that he intended to introduce an Islamic constitution throughout the country. The Sudan People's Liberation Army (SPLA) began rebel attacks in earnest in 1983.

On a morning in March 1987, armed Arab Rizeigat mobile militiamen attacked Dinka worshiping in their church. That night, many more of the same tribe's houses were looted and torched and people were killed before government officials summoned the Dinka over loudspeakers to group at the railway station near Hillat Sikka Hadid, where they were told a train was ready to transport them away from

Daien. That evening, assailants lit grass mats and thatch they pulled from the roofs of surrounding houses. Pushing the burning material under the train wagons and through the open windows of the compartments loaded with Dinka, the attackers suffocated and scorched those captive in the carriages. When they struggled to escape, they were robbed before they were bludgeoned to death or killed with knives and guns.

"About 6 p.m. the massacre stopped," the *Sudan Times* reported on 17 July 1987. "The Rizeigat left."

Government officials then took to the streets with loudspeakers and requested the Dinka in hiding to head toward the railway station. "At 7 p.m. the first train started to move toward Nyala. It pulled only one wagon, a wooden one . . ."

Following the massacre, dry grass was spread about the wagons and fuel poured on the bodies of the dead in an attempt to burn the terrible evidence. The ashes and bodies of those who died in the bloodbath were buried in two huge mass graves, one close to the railway station and the other at the Daien town graveyard. The killing had almost certainly been premeditated.

Articles in the *Sudan Times* continued to express concerns that the Dinka tribe lost thousands before being shifted to Kass in central southern Darfur, where they faced starvation and terrible deprivation (and still do).

Assuming the information reported in the daily papers to be accurate, we scanned the newspaper regularly for news and views. The local media had extensive influence with the country's populace and, of course, the government was concerned with what was reported. Twenty -five of the thirty Sudanese newspapers were printed in Khartoum, and Bona Malwal, the editor of the *Sudan Times,* stated that, of those, 80 percent were owned by the Muslim Brotherhood, and all but two were printed in Arabic.

Chapter XXII
To Touch the Sun

WIND WHIPPED LITTER in front of a surreal sun that showed blood red behind a cloud of brown dust. Forced to stop the car, I felt uneasy with the ferocity of the *haboob* that howled around the Land Rover in which Jhorna, Sanjeeb (who was home for vacation) and I were seated. Shrouded people groped past us in the overwhelming dust. Suddenly, drops of rain blotched patterns on the windshield. Welcoming the sweet smell of wet earth, we followed a dirt lane that led to the stables owned by our Egyptian friend. Laughing children ran behind the car until I found a place to park under the shade of a tree. Grimacing at the reflections of their dirty faces in the rear view mirrors, the kids followed us to where we found Jessica stroking the handsome Arabian horse she had galloped along the Nile. The horse tossed its beautiful head, startling the children, who ran in every direction.

"We won't be on safari long, will we?" Jessica inquired as she closed the stable door and affectionately patted the horse.

The next day we packed the car with our camping equipment and drove down the road to where uniformed men motioned for Cole to stop the Land Rover. Nearby, passengers milled about a small tent, waiting for their bus driver to complete the procedures necessary for them to journey on. The military guard examined our travel documents carefully before he allowed us to join the other vehicles traveling southeast on the new tarmac road. Buses and lorries loaded to capacity bore down on us at high speed as Cole drove us toward Wad Medani and the vast irrigated cotton fields known as the Gezira scheme. Farther on, evening light transformed the fields of *dura* to warm gold. Cole turned off the main road and followed a small track that led toward a distant hillock. Eager to set up camp, our two eldest children, who had returned from boarding school, called to Jhorna to help them put up the tents.

Clouds of insects rose from the fields the next morning as we drove on toward El Gedaref. Silhouettes of *jebels* burned in the pale sun a few miles from the Ethiopian border, where one million Ethiopian refugees resided. Resplendent in their red-and-black appliquéd skirts, Rashaida women, whose attire reminded me of the Lombardis I had grown up with in Andhra Pradesh, walked near the Kassala souk, where I bargained for local jewelry and clay *jebanas*.

Down the road, thousands of camels were gathered around a muddy pool, drinking. Surging and restless, Sudan's export to Saudi Arabia waited impatiently for the herdsmen to pour salt into the murky water trapped behind a small dike of sand for the animals to drink. Other men filled great leather skins with water to overflowing. Suddenly, with a loud shout, a man abandoned his herd and raced after a shaggy male that fled bellowing toward the windswept plains. We watched the retreating camel, the herdsman in hot pursuit, until the incident became a mirage. That night, when stars patterned constellations in the indigo sky and moonlight outlined rocks sculpted by the elements, Cole and I wondered at the rifle shots we heard in the area, as it was known for banditry.

Fine sand traced the wiper blades the next morning as Cole followed the road that led toward the purple shadows of the Red Sea Hills, the home of the Beja. Traveling on to Sinkat and the shores of Suakin, the oldest Arab port on the African side of the green-blue Red Sea, we stopped near crumbling ruins that spoke of bygone days, when slaves were traded for silk. We shared sweet, pink grapefruit with beautiful flush-cheeked local children, who demanded baksheesh. Seventy miles north of Port Sudan, we set up camp at the end of a small peninsula, not far from local fishermen who cast their nets and supplied us with filleted fish for our meals. By day, we moved with the slow rhythm of the ebb and flow of tides. And when darkness fell, the children crept under their mosquito nets, which shivered softly in the night breeze. Lulled by the gentle lap of waves accented by the light of the moon, they soon fell asleep.

Seven sun-baked days later, we journeyed on toward Dinder Reserve, where thousands of guinea fowl scattered through acacia trees with red and white bark. Rabbits scampered across the road the next day as we returned to Khartoum, already making plans for our next safari.

Some weeks later, on a journey to Musawwarat, Cole branched north from the paved road. Following the tire marks left by other vehicles, he drove through the desolate desert terrain alongside the railway line, to get to the ruins of the Kushites. Eyeing the compass from time to time, Cole kept a steady foot on the pedal as we sped swiftly over the sand toward mirages and undulating sand dunes. Experienced in handling four-wheel-drive vehicles in difficult situations, he shifted into low gear and coaxed the Cross Country through the deep sand of an old riverbed. The children leaned closer to the front seat and urged him to continue the story of how four disoriented travelers had driven in wide circles until they lost track of time and direction.

"Their dried bodies were like mummies when they were found in a Land Rover years later by a wandering nomad," Cole said, squinting at the sun through round sunglasses. And although the children trusted their papa's sense of direction, they shifted uncomfortably in their sticky seats and were relieved to see signs of humanity and a cluster of huts ahead.

Startled to see us, a little shepherd girl fled and was called back by a man who wielded a stick behind a small herd of hungry goats.

"*Salaam aleikum,*" Cole said, slowing the dust-covered Land Rover, and inquiring about the direction to Musawwarat. Withered as the thorny bushes that struggled to survive in the barren landscape, the man motioned for us to continue in the general direction we were going. "*Insha'allah,*" he muttered, no doubt willing us to reach our destination as we left him silhouetted against the sky, the hot wind flapping his faded *jelabiyya* against his bony legs.

"Perhaps we should have followed the main road to Meroe and proceeded from there," Cole mumbled, as we strained our eyes to search for the outlines of hills. But then, when the shades of the ocher landscape softened in the last rays of the sun, we saw rock pillars silhouetted near ruins of stone supporting drifts of rippling sand. Anxious to leave the confines of the car, we instinctively trod with respect on the timeless scene.

"Far to the south of Egypt along the Nile's great meander, lies the heartland of an ancient African kingdom of which few have heard and much remains to be learned," Timothy Kendall had written in the *National Geographic*. "The Egyptians, who conquered it in the 16th century B.C. and ruled and exploited it for 500 years, called it Kush—'wretched Kush.' For they both loathed and dreaded 'this land of whirring wings . . . [with] a people tall and smooth skinned . . . feared far and wide, an aggressive nation of strange speech,' as the biblical prophet Isaiah would describe. Pharaoh Tutankhamen had images of the black-skinned Kushites embroidered on the soles of his sandals and carved on his footstools, so he could perpetually trample on them. . . ."

When the Egyptianized Kushite kings conquered Egypt in the eighth century BC, they established a vast riverine empire extending from the confluence of the Blue and White Niles to the Mediterranean.

That night, the children exclaimed that the stars were so close they could reach out and touch them. Bedding down on the warm sand, we searched for constellations.

A breeze like a whisper woke us at dawn and the children scrambled atop the solitary hill nearby, followed more slowly by Cole and me. Taking care not to step on little creatures that scuttled under dry brush, we shaded our eyes and looked with awe on the endless vast wasteland of sand—once fertile, as evidenced by large pieces of petrified wood we found bleached by the sun.

Omdurman

Sheep with matted wool marked with red dye were huddled near a fence, waiting to be thrown mercilessly into the boot of a car or tied to the back of a pickup to be driven to slaughter. Abdul Manjit pushed the brake impatiently as we passed a donkey cart laden with goods to be sold at Suq al-Kabir. Mud walls obstructed the view of swept courtyards and minarets rising from mosques on the sides of Omdurman's dusty streets. Farther on, under the shade of large trees, Abdul Manjit parked near a white wall outside the Prime Minister's residence. Well-known to the inhabitants of the compound because he had driven Prime Minister Sadiq al-Mahdi's father (the grandson of Mohammed Ahmad al-Mahdi), Abdul Manjit led us to the back entrance of the house, where we were told to sit on the verandah near a canopy under which there were many *angarebs*.

Soon after, an old man escorted Jane Campbell, Cole's deputy, my mother, who was visiting us, and me up the stairs to where Seyyida Sarah, the Prime Minister's second wife, showed us to a small, comfortable room. Offering us cakes and tea, the warm and hospitable Sarah favored my gracious mother, and was curious to know about the many years she had spent in India. She then conversed easily about the various activities she was involved with, and told us she was pleased to have completed her master's degree in sociology. In the midst of telling us of her imprisonment during Jaafar Nimeiri's rule, she paused to acknowledge her husband, who entered the room.

The great-grandson of the Mahdi was distinguished in his flowing white robes. After greeting us, Sadiq al-Mahdi inquired as to our comfort and exchanged a few pleasantries before he excused himself, followed by Sarah. She returned from the adjoining room with a brass kohl receptacle, which she presented to me as a gift. Demonstrating the technique Mother and I were familiar with from India, she told us she used the soot made from burned cotton mixed with olive oil to accentuate her eyes.

The sound of the muezzin calling people to prayer came too quickly, reminding us it was time to leave. Burnt sienna colored the western sky as we waved goodbye to our new friend and drove out the gate.

Two years after our late afternoon tea with Sarah, in June 1989, Brigadier Omar Hassan Ahmed al-Bashir took over Sadiq's position in a bloodless military coup d'état, and Sarah and her husband were put under house arrest.

Of Henna, Halawa and Brutal Traditions

Shahwa lifted the *jebana* of coffee from the woven grass coil and poured a little of the strong brew laced with cardamom into a small clay cup. In contrast to other Sudanese women, who peered through black veils and swathed themselves in black robes, Shahwa covered her head with a gauzy white *tobe* that slipped seductively to her shoulders.

Spooning sugar into my cup, she inquired of Lula, "How much sugar?" as the three of us sat wilted in the Khartoum heat under the shade of an Al Sudaniya hut. Accepting my compliments on the ornate pattern that stained her hands, Lula offered to take the girls and me to Omdurman for what she termed "our tradition."

On a later day, Lula led the girls and me through a courtyard to a room where her friend motioned for us to sit on *angarebs* pushed against the wall. Expertly squeezing thick dark paste from a plastic funnel into patterns on the girls' hands and wrists (although she noted that in Sudanese tradition, this was reserved only for married women), the woman explained in halting English that the paste was made from the crushed, dried leaves of the henna plant mixed with a little water and dye. Lula, who was a regular client, carried on a lively conversation with her friends, one of whom drew intricate designs on my hands and feet. Near me, another woman was having her slim feet and ankles painted with delicate patterns that would be partially concealed by the skirt of the tunic she wore. (Much later, a Bangladeshi friend told me she thought it disrespectful for women to have henna painted on their feet, as the prophet Mohammed had worn henna in his beard.)

We were then told to sun-dry the patterns on our hands, before the paste was washed off and we held our hands under a blanket over sandalwood coals and the thick smoke of fragrant resins. Delighted at the rich brown patterns that emerged on our hands and feet, the women complimented the girls and told them the henna would last for at least ten days if they washed carefully. Lula returned our grateful hugs, and I commented on the wonderful fragrance she wore.

"It's *dilkah*. I shall give you some," she said, with the easy generosity I had come to admire among my Sudanese friends. On the way to her house to get some, Lula explained how she made the scent by mixing *dura* with water in a wooden bowl that she covered with a blanket and smoked for two or three days. "After that, I mix cloves, spices and sandalwood powder with the flour and smoke it again, then add some of my favorite perfume and store it until I need it." That night, I combined some of the mixture Lula had given me with oil and spread the sensual scent that was said to keep one's skin moist against harsh weather conditions on my arms and legs, as Sudanese women had done since long before Biblical times.

Another day, I joined a friend in her apartment adjacent to ours for *fatuur*. Dipping a piece of *kisra* made from ground millet that had been fermented and cooked into thin bread in a bowl of *mulah* stew, I ate the tasty food while Leila, who earned her keep by walking from house to house on heavy, arthritic legs to remove her client's unwanted hair—made *halawa* with the juice of two lemons, one cup of sugar and a little water she boiled then cooled.

Dipping her hands in water, Leila pulled the *halawa* until it stretched into silky ropes, then spread the taffy on Mona's shapely leg, which she then pulled off to remove the hair.

"It hurts a little, but *halawa* is good for the skin," Mona told me as I winced. "Before I was married," she continued, "I had my underarm, facial and even my pubic hair removed in order to make my skin soft and smooth." The wife of our landlord, Mona had three children by her husband. He had previously been married to an English girl who he claimed had shown him the full gamut of Western decadence.

Karim had returned to the Sudan with one daughter and now practiced the Muslim custom of reverence to Allah and kept his family under tight surveillance. But Mona was content with her routine. She kept her hair in curlers and looked after her children until her husband returned home for late lunch. While Mona rubbed oil on her smooth legs, Leila inspected my arms. She told me frankly that the hair looked unsightly, and forced me to undergo the *halawa* treatment. I was further chastened when she said she must "repair" my cracked heels. Shaking her head at what she warned was neglect, she said she would return in a week to remove the calluses from my feet—something she did deftly, albeit painfully, with a razor blade.

"Customs, rituals . . . grandmothers in the Sudan have so much influence," Ellen Ismail, an Al Sudaniya colleague and German friend, who had been married to a Sudanese for twenty-five years, told me one day. "In fact, grandmothers are often the ones who insist on traditional rituals and female circumcision." Intrigued to hear more about female genital mutilation (FGM), I joined a number of women in a private home to hear a Sudanese gynecologist talk about FGM and the dehumanizing violation of children's rights. Informing us that the practice had been banned in the Sudan in 1946 and that a small minority of educated urban women in the Sudan condemned female circumcision, the doctor leading the group told us FGM was still widely practiced.

"These days, a few six- and seven-year-old girls are circumcised under sterile conditions in clinics," she admitted. "But village women operate on the girls without anesthetics. They cut away the clitoris and sometimes the inner and outer labia of the child with a piece of glass, scissors or razor blade. They then fasten the child's wound with unhygienic threads, horsehair or even thorns, before they bind the girl's legs together for up to a month to allow scar tissues to grow around a small opening for menstrual blood and urine to flow through." *Flow?*

"FGM causes numerous complications," the gynecologist continued telling the group of women I was with, none of whom had endured the cut. She concluded by saying, "Hemorrhaging, painful and prolonged menses, urine retention or repeated urinary tract and vaginal infections, incontinence, psychological trauma, sexual difficulties from tearing and natal problems are all potential problems resulting from mutilation. And if the scar tissue forms too tightly around the opening of the vagina, re-cutting must occur for consummation of marriage and again during childbirth."

Sickened, I listened to further discussion among the women, one of who claimed the practice was said to have begun in Africa before Christ and that female circumcision was thought to prevent sexual play. "The girls are stitched up and remain this way until their wedding night, or their husbands might reject them," she said. "As well, men prefer intercourse with a small tight opening."

"If the operation of circumcision is to stop my Sudanese friends from having boyfriends, it really doesn't work," my young Ethiopian friend told me with a disdainful laugh. "Anyway, when the girls want to marry, they just have themselves stitched up again," she added, with a toss of her curly head.

Prepared to leave the host country they had taken refuge in, the refugees we came in contact with hoped it would only be a matter of time before they found ways and means to leave the Sudan. Lines of hopeful Ethiopians who had walked over the border, many of whom had lost their families and property, waited for months to enter their application for foreign visas. In the meantime, their hope was to find work to support themselves—such as soft-spoken Sam, who made leather sandals he marketed at Al Sudaniya, and Joseph, who wove crosses and belts from strips of leather.

Children Who Run Under the Sun

Ahmed leapt up the stairs to the American Embassy, his long black legs lean and bare below his blue shorts. One of fifteen boys associated with Street Kids Incorporated (SKI), he carried a fluorescent orange delivery bag over one shoulder.

Founded by Peter Dalglish, SKI challenged the prejudices against street children. Once destined for life on the streets, the boys wore red caps printed with their logo and wheeled bicycles through Khartoum's traffic snarls, offering courier service to agencies and businesses. The children were convinced they needed to learn to read and write in order to ensure they delivered letters to the correct addresses. With their own earnings, they paid for a tutor to teach them Arabic and English.

The *shamassa*, children who run under the sun, who numbered about ten thousand in 1988, survived on the streets in Khartoum by rummaging for food in the city's rubbish heaps near the main market. The street children gathered outside Blanka's ice cream shop every evening. There they received leftover ice cream and sometimes bread and milk from our Yugoslavian friend, who had lived in the Sudan with her husband for thirty years.

"The best I can do for these children is to give them food," Blanka remarked to me one day. And when the boys' medical complaints were beyond "Mama Blanka's" advice, her friend, Dr. Ali Abdul Karim, treated the boys for ailments such as syphilis, bilharzia, wounds, anemia, tuberculosis and other side effects of abuse, neglect and malnutrition.

Blanka was sympathetic to the plight of her "lost boys," and told me the kids suffered spasms and disorientation from sniffing glue, benzene or burned rubber.

"They practice the habit," she said caringly, "because sniffing temporarily takes away their hunger pangs and leaves them pleasantly high. How can you blame the children for turning to prostitution and thieving to obtain money for their fixes? But you know," she said, after a long pause, "I believe no human alive on this earth is useless."

Two boys who regularly visited her shop had traveled from the south of Sudan to Khartoum. Their mother had put them on a lorry to go north in the hopes of them finding food. "The elder of two boys assured me that he was strong and willing to do anything," Blanka told me, "as long as he could get food and keep his younger brother with him. Unfortunately, my husband's family already had enough small boys helping with errands and I had nothing for them to do. But," she added, "self-dignity is important for the boys. When they are reduced to begging, they feel useless and start to behave in this way. But they are not useless. I have never met one of them who, given a chance, did not jump at an opportunity." (This was something I found to be true when, much later, I worked with street children in Kenya.)

SABAH, the beautiful word for "morning" in Arabic, offered many boys support and an opportunity they otherwise would have missed. Munir al-Safi administered SABAH, which provided street children with a shower, basic first aid, a morning meal and a counselor to provide attention and information if that was what was needed. Afternoons and weekends, Peter Dalglish organized training for Khartoum's *shamassa* at the Belgian Technical School when the classrooms were not in use. Following careful selection, the boys were given literacy classes and divided into groups for theoretical and practical training in skills such as mechanics, electricity, carpentry and metalwork. Although at first reluctant, the Belgian school staff soon recognized the boys were more eager than the regular students in courses that took nine months to complete. The boys graduated with a technical certificate from the Ministry of Education with a success rate of nearly 100 percent.

Human Rights Violations

One day, as Cole folded an extra shirt and pair of trousers into his bag to travel to Wau, I lamented, "It's only been a week since Eddie Folman's 404 was shot down in southern Sudan."

"Less than a week," Cole reminded me. "But Daniel's a good pilot. He checks and double-checks and knows his options if anything goes wrong. Remember he worked as a bush pilot in the military conflict in the Belgium Congo before he came to the Sudan. Anyway, their lives are in danger," he said, referring to the MSF emergency doctor and nurse he had placed there six months earlier. Turning to give me a hug and a kiss, Cole went out the door that January morning in 1988, to evacuate the medical workers from Wau.

One week before, soon after flying out of Juba on a four-hour flight to Khartoum, Cole looked down at the nondescript bush country and remarked that he saw red flashing lights on the ground. "You should have seen Daniel shoot out of the pilot's seat," Cole told me later. Looking out of his window to see another three flashes, Daniel popped back into his seat and pushed the controls of the engines to full throttle to increase the plane's altitude. Thinking the flashes looked like heat-seeking surface-to-air missiles, Daniel told Cole to keep looking and call out if he saw them again. Meanwhile, he lifted the plane to seventeen thousand feet—a height he hoped the missiles could not reach.

A few minutes later, Cole saw what he later described as "a beautiful configuration of bursts of smoke." Seconds later, explosions surrounded the smoke, and Cole told Daniel that warheads had exploded two or three thousand feet below the Twin Otter. "Daniel was very tense," Cole remembered later. "He had beads of perspiration covering his forehead even though the temperature at the altitude we were flying was below freezing."

After the SPLA shot down a Fokker Friendship plane over

Malakal, killing all sixty people on board, all commercial flights to the south were stopped. Soon afterward, in 1986, ICRC evacuated their medical team from the western part of southern Sudan and suspended their operations in Wau. The Sudanese national officers who kept a UNICEF presence at the small office there radioed Cole that the SPLA had surrounded the government-controlled garrison town and were shooting at supply planes from the north. "Convoys trying to leave town are also being shot at and the town is being starved out," the national officer warned.

Concerned that medical care was nonexistent, Cole conferred with the director of Médecins Sans Frontières, Holland and told Jacques de Milliano that if he could find the right medical people, he would get permission from the SPLA in Addis and the military government to fly the UNICEF Twin Otter to the south.

A week or so later, Jacques reported that a doctor and nurse were willing to go to Wau. (Much as war correspondents associate themselves with risky conflicts, it is not uncommon for emergency personnel who live from one disaster to another to become relief junkies.)

Before dawn, Daniel took off with Cole and the medical personnel, who had been supplied with medicine kits and two-way radios. They would fly via El Obeid in order to refuel for their return flight. Arriving safely in Wau, they were met by a UNICEF staff member and government officials, who took them on an immediate tour of the hospital. It was filled to overflowing with outpatients suffering from malnutrition, diarrhea and measles. When Cole and Daniel were ready to depart, dozens of people determined to leave Wau followed them to the plane, but were told that only four could fly to Khartoum. They then took off on the little-used east-west runway close to town, rather than using the main runway they had landed on. Checking his controls, Daniel put the big turbo-prop engines to full revs before he lifted the high-winged aircraft off the ground. Once airborne, Daniel turned sharply north, away from where he thought SPLA soldiers might be located with rockets or missiles . . . but not soon enough to avoid the tremendous *thud* they heard hit the bottom of the plane.

"The plane shuddered at about five or six hundred feet above the ground," Cole told me later. "But we continued to climb to fourteen thousand feet."

Then, although they had already refueled, Daniel made an unscheduled stop in El Obeid to assess the possible damage caused to the plane. Fortunately, they landed without incident and reached Khartoum just before sunset. The doctor later told Cole over the two-way radio that when they had taken off, there had been an explosion on the ground behind the track of the plane. "But the military people with me just shrugged their shoulders and said we should return to the safety of the town," he reported. Daniel concluded that one of the plane's balloon tires (made to land in the bush) had been hit by a rocket-propelled grenade. He guessed that the grenade had ricocheted off one of the big tires.

Because of the above incident and increased SPLA activity, the planned monthly flights to Wau became impossible. Cole evacuated the medical team nine months later.

Soon after, the MSF doctor shared an evening meal with us. Pushing the food around his plate, he shook his head and said. "There were at least fifty thousand in the township of Wau who were starving. There has been so much killing." The doctor was no amateur. Skilled as a marksman, he was not easily intimidated. He had been part of the Dutch army's anti-terrorist squad, and he formed a disciplined militia in Wau, with Dinka soldiers who acted as bodyguards and patrolled the hospital grounds. Anticipating the conversation that would follow, I excused the children from the table and listened with dismay as he spoke on about unspeakable atrocities. Of massacres of civilians, mutilated bodies and children impaled on spears, nailed to trees, of rape . . . sickened, I pushed back my chair.

The doctor's concern was for those remaining in Wau. If the international media was alerted to generate publicity, he said, "the accusation of genocide would cause instant death for the people there."

It seemed the Sudan government and its armed forces were operating a carefully constructed strategy; one that ignored violations of the tribal militias against southern civilians, in order to destroy any possibility of support or assistance for the SPLA. This phase of the conflict continued hardening hearts and attitudes for six years. The army controlled less and less territory in the south and the Arabs feared the war would extend to the north. More and more displaced southerners arrived in Khartoum, where they were charged taxes for living on the city's rubbish heap.

Prayers, the Nile and the Displaced

Rotating fans offered little relief from the excessive heat and we shifted uncomfortably in the pews of the cathedral we normally attended on Friday—the Muslim day of prayer. Pausing a moment before he spoke, the speaker of the evening adjusted his papers on the pulpit before he pleaded with the congregation to practice Christ's teaching.

"Love one another," he said. "We will go nowhere unless we love God and one another."

I heard his tearful voice speak on about the need for reconciliation and dialogue between those who God created—had made in His image.

There were those, however, who bred suspicion among this and other congregations. And one day, during a service held in the Christian sanctuary, a tribal feud caused disagreement and three people were seriously injured in the confrontation, which caused much material damage. After that gross misrepresentation of Christ's love, God's house became a vacant building, as the government kept the doors to the cathedral locked.

Although repeatedly warned to take extra precautions, and more so after a bomb threat at the Hilton Hotel, we regularly frequented what had been Kitchener's gunboat, which was now the Boat Club, moored under *labakh* trees that shaded the Nile. The *Menelik*'s stairs led to the water, where our small inflatable boat bobbed near the buoy to which it was tied. If the two older children were home for the holidays, all three donned water skis to skim to Omdum Island through water that swelled waves toward those standing on the banks. On occasion, Cole guided the boat through the muddy waters of the White Nile, which merged with the Blue near the bridge that crossed to Omdurman.

The White Nile originates from where Cole used to cast his fishing line in Uganda near the foaming Ripon Falls of Lake Victoria. Flowing 975 miles from Lake Tana, the Blue Nile normally begins to rise in March and April when the rains fall in the highlands of Ethiopia. But in August 1988, unusually heavy rains added to the already swollen waters of the Nile, causing the worst floods in the twentieth century. Over a period of thirteen hours, eight inches of torrential rain lashed Khartoum. Mud structures collapsed and more than a million people lost their homes in water that was over three feet high in places. The city was without electric power for five days and many areas were without fresh drinking water as the rising Nile continued its destructive course. Bursting its banks and inundating large areas of northern and central Sudan, the swollen river caused tens of thousands of people to move their belongings in order to escape the floods.

"The electricity, water, telephones, roads and rail have all been cut in Khartoum," *The Independent* reported on 29 August 1988. Thereafter, international correspondents focused their attention on the refugee situation as well as the displaced southerners, whose pathetic structures on the outskirts of town had washed away. In the south of the Sudan, Equatoria, Upper Nile and Bahr al-Ghazal experienced floods that destroyed homes. Grazing land was submerged in areas where people already well accustomed to suffering were displaced because of the civil war. "Six million people were caught between opposing armies and armed bandits," correspondents reported, ". . . survivors who experienced looting and plundering of their households, herds, local dispensaries and schools."

Chapter XXIII Sudan's Misery and a Charged Atmosphere

THE SUDANESE ARMY was quick to capitalize on traditional animosity and provided food, fuel and ammunition to the militia they armed to fight a "proxy" war. Meanwhile, the SPLA chased local inhabitants away from their settlements and fields and ambushed relief convoys from regional capitals to outlying small garrison towns, often burning their crops, looting granaries and destroying the food they could not carry away.

Struggling citizens paid exorbitantly high prices for basic commodities, while others took advantage of food shortages to amass personal wealth. In mid-1987, the price of a bag of sorghum used for porridge that sold for 50 pounds in the north traded for 2000 pounds in Wau. The price of a single teaspoon of sugar in early 1989 was one pound.

Full-scale famine resulted. "Standing between food and the people," James Wilde wrote in *Time*, "are 58,000 government troops and 30,000 rebels. On both sides, the terrible weapon is increasingly food, not bullets."

Around the same time, on a flight from Paris to New York, the executive director of UNICEF opened the centerfold of a newspaper and was horrified to see a picture of reed-thin southern Sudanese children photographed beside corpses. "We've got to do something," Jim fervently told Cole over the telephone, then asked him to report to New York for further discussions.

The head of the African desk and program division, as well as others at headquarters, opposed taking the kinds of risks involved to get the relief operations going. However, Cole received Jim's permission, "even if there's no way to do it officially," to obtain a letter in writing from the prime minister, an agreement with John Garang and SPLM and a memorandum of understanding together with the International Committee for the Red Cross, ICRC (who already had permission to work in the south) to authorize UNICEF to work in the SPLA-controlled area in southern Sudan.

Discouraged when one attempt after another failed to coax the prime minister to begin negotiations with the SPLA, Cole paced the floor and told me Sadiq's response was hesitant. Finally, the prime minister and the minister of defense came to an agreement allowing Cole to initiate a dozen flights to southern Sudan to deliver relief. Following meetings with the SPLM high command, an agreement was reached, centered on informed consent. The UNICEF Twin Otter would be allowed to fly to three northern-controlled garrison towns and three SPLA-held locations.

Encouraged, Cole informed me the flights would originate from Khartoum and would begin to fly the first week of December and thereafter the first week of each month. But just as plans were falling into place, the SPLA captured Kurumuk (a town near the eastern border of Ethiopia), with the result that the minister of defense refused to sign the final one-and-a-half-page army draft.

In the meantime, Cole set up a cross-border operation in June 1988 to deliver food and medicine to the south of Sudan from Nairobi. This clandestine program, he hoped, would help coordinate cooperation between the SPLA/M, the NGOs and the civilian population in order to protect the rights of children, as well as provide a chance for relief organizations to have access to the interior of southern Sudan for the first time since the outbreak of the civil war in 1983.

Rising at dawn the morning he was to meet with the prime minister, Cole slipped on a pair of trousers and a shirt and left our flat to walk the dusty streets of the neighborhood, his mind churning. He returned home to find the bloodstained feathers of a pigeon strewn on the staircase. Convinced it was an omen, he carried the wounded bird up the stairs and placed it under the eaves of the roof.

At their meeting, the Prime Minister ignored Cole's extended hand. He focused his eyes straight ahead and said, "I have come to learn you have transgressed the boundaries of the Sudan without the consent of the government. You are helping the enemy of democracy. Thereby you are prolonging the war and contributing to the death of my soldiers."

Sick with disappointment, Cole left the office. Some days later, he flew to Addis Ababa, Ethiopia, without the approval of the Sudanese government. There, members of the SPLM taunted him for having trusted the northern government in the first place.

Having heard about Cole's attempt to help the southerners, a friend and former Oxfam colleague of Cole's, Betty Wetzel, wrote a letter to President George W. Bush, saying:

Dear Mr. President,

I wish to protest our financing of Sudan's war against its own people. Given our straitened circumstances at home, which do not allow us to house or educate our own people, much less provide them with minimal health care or tend to our rotting cities and crumbling public services, this expenditure of millions to finance a corrupt regime is not only foolish. It is immoral.

The recent article "Famine," by Raymond Bonner in New Yorker, is truly chilling. The information that Sadiq al-Mahdi has prevented reputable international aid organizations from providing food and medicine to the starving people of south Sudan is almost unbelievable.

The fact that he ordered Cole Dodge, Unicef director, out of the country convinces me of his corruption. I worked with Cole Dodge for several years at Oxfam-America and have followed his humanitarian career in the third world since that time. I know that he is a man of utmost integrity and compassion. His ouster should have been protested by the U.S. and the whole international community. His ouster was a confession of guilt . . ."

Soon after, the Sudanese government expelled four Western donors working with the war-inflicted famine in the south. Fearing the same fate, and frustrated with their inability to provide aid to the south, other humanitarian organizations turned their annoyance toward each other. The atmosphere became charged and tension brewed between NGOs and bilateral donors.

Disgruntled that UNICEF was flying the only agency aircraft from north to south, four northern embassies complained to UN headquarters in New York about UNICEF's activities. "Sadiq al-Mahdi was democratically elected in 1986," the diplomats said. Labeling the SPLM as an illegitimate movement, they criticized the SPLM for not participating in the elections and as much as said that they were not willing to give any assistance whatsoever to southern Sudan in SPLM/A territory.

Hoping to draw donor and media attention to medicines and food being delivered by train to the stricken area, and to facilitate NGO personnel to work with those affected, Cole flew south with Gary Striker from CNN, a BBC reporter, aid officials and two ambassadors. Scrambling off the plane, the journalists set up their photography equipment and took pictures of the relief goods being loaded on the train. But it was not until 3 a.m., and after Cole had paid extra money to the reluctant engineer to purchase a barrel of oil for the gearbox of the engine, that the train pulled down the track toward El Muglad. The team then flew ahead to El Meriam, near to the north-south border.

Meanwhile, newspaper headlines continued to cry: "Starvation!"; "Time Is Running Out"; "The Vulnerable in the Sudan Are All Dead." But still the Arab field commanders failed to report mass deaths.

While the SPLA blew up convoys and threatened to shoot down relief planes, the journalist Hindawi reported, "Soldiers of fortune flew Hercules C-130s to the south carrying arms and ammunition and returned loaded with ivory and ebony."

Both sides of the civil war continued to use food as a weapon. Caught between the government and the rebels, "hundreds, if not thousands of civilians die each day," the editorial of *Sudan Times* reported on 16 January 1989. "All economic activity has ceased. The traditional cultural system of the nomadic population is threatened because there are not enough cows left for marriage or for settling disputes. But worse, southern Sudan is in danger of losing its people."

Left to wander through bush land planted with mines, innocent victims abandoned their crops and fled from their homes to seek refuge in overcrowded towns. Abyei's population leapt from ten thousand to forty thousand despondent, listless people with wasted limbs and sunken eyes. By night, packs of hyenas crept into the sprawling camp to drag away emaciated corpses.

In time, and after a great deal of negotiating, voluntary agencies working with the UN were able to provide food and medical assistance to El Meriam and Abyei, all the while accusing the government of Sudan of outrageous human rights violations and blocking relief efforts.

"I Have It"

At the controls one morning, Jim Gaunt took off with Cole and ICRC personnel from Lokichoggio in northwestern Kenya near the border of Ethiopia and the Sudan. "Gaunt landed the plane somewhere in the bush," Cole told me later, "but it was well worth the four- or five-kilometer walk to get John Garang to sign a handwritten letter allowing UNICEF permission to work in the south."

Pleased to have the letter stuffed in his pocket, Cole and the others returned to find Gaunt pacing the length of the runway. "It's late," Gaunt said shortly, putting the engines at full rev to the end of the rough *murram* strip. Releasing the brakes, Jim gained speed and lifted the vibrating plane over the tops of trees toward Lokichoggio, where he landed toward the setting sun.

Mike Grennell, who we'd worked with in Uganda, was not so fortunate. Last seen on the same day repairing his plane, he crashed the Cessna 337 aircraft he was flying, and both pilots were killed.

That evening, Cole and the ICRC delegate drafted a handwritten memorandum of understanding between UNICEF and ICRC. The next morning, Cole flew to Nairobi to have the memorandum typed and initialed by ICRC. Together with Cole, they agreed on how they would work together in the Sudan. Communicating the good news to headquarters, Cole flew back to Khartoum via Addis.

Meanwhile, in far-off South India, I said goodbye to our children, who I had been visiting for a month at the boarding school they attended in Kodaikanal. Boarding a plane in Coimbatore for Mumbai, I flew via Addis to meet Cole en route back to Khartoum. I was so eager to see him that I took the airport stairs two at a time to reach the first floor. I was startled when Cole pushed me behind a pillar before he gave me a hug.

"Don't acknowledge that you know Jane," he warned in a low voice.

"Why, what's up?" I asked. It was then I noticed how pale and haggard he looked. With growing apprehension, I listened to Cole relate the tension-filled drama that had transpired during the weeks of our separation. He ended with, "I will very likely be declared *persona non grata* when we arrive in Khartoum," as we walked together to the airport lounge.

Earlier, the Sudan government had expelled Winston Prattley, the head of UNDP and Special Representative of the Secretary General for Relief Operations, with the rank of Assistant Secretary General, because of discussions he'd had with the SPLA over emergency supplies. Fearful now that she also would not be allowed to reenter the Sudan, Jane, Cole's deputy and the wife of the British High Commissioner, had flown to Addis to warn Cole he might be arrested when he returned to Khartoum. She sat a little distance apart from us in the airport, flipping absentmindedly through the pages of a book she obviously was not reading.

So filled with apprehension and misgivings was I that for once I was unaffected by turbulent skies that delayed our plane's approach to Khartoum. Worried he would be unable to deplane, Cole had written a list of priorities, which included the sale of our personal goods. I slipped the list into my coat pocket and peered anxiously from the window as the plane came to a halt on the tarmac.

"There's no sign of military or police," I said quietly to Cole. "Only a lot of hot air . . . and Abdul Manjit." Abdul stood at the foot of the stairs and extended his hand. "Did you have a good flight?" he asked.

"*Alhamdulillah*."

A few days later, the Twin Otter's engine droned over the sun-bleached canvas hazy with dust. Pressing my forehead to the window, I watched the track far below wind toward baked houses and the town of El Obeid. Ahead of me, Cole occupied the seat next to Daniel in the cockpit. He was always ready to take the controls himself, having soloed in four days and obtained a license after a week's course in 1982.

We were landing in El Obeid to pick up those who would fly on to Abyei. The tarmac reflected the hot sun as I disembarked behind three Oxfam and Save the Children nurses. The women were blatantly uncommunicative. They were resentful of Cole, who suggested they wait for the second flight as he had sent reporters on the plane they had expected to fly on to cover the relief operation. Later, with accumulated frustration, the nurses wrote to their headquarters, accusing UNICEF of delaying aid and obstructing their program.

Shortly thereafter, with the blessing of James P. Grant, Cole flew from New York back to the Sudan via London to hand-deliver a letter to SCF, who, although they had known Cole for years, had accused him of following the wrong policy, and one that was detrimental to the children of the Sudan. Hearing this, I stifled a groan, for no one, in my estimation, had worked harder for affected children through the years. In London, Cole signed the first debt relief for development loan with Bankers Trust and called Jim (whose wife Ethel had recently died) to tell him, "I have two signed letters."

On 2 January, 1989, just after our two eldest children had returned to boarding school in India, Cole was again on the doorstep of the Prime Minister. This time, Sadiq al-Mahdi received Cole warmly and told him to draft out what it was that he wanted.

A week went by, and nothing happened. In the interim, the Secretary General of the UN, Javier Pérez de Cuéllar, called Jim Grant to his office and told him he had received complaints about UNICEF in Khartoum. Since Jim was going to be in Cairo for a day, he suggested Cole fly to meet with him there.

"I have two letters signed by the ICRC and SPLM," Cole told Jim. "But we still need a third: a letter from Sadiq."

"What are the chances?" Jim asked.

Informing him he hoped it would just be a matter of time, Cole added there were many rumors flying around Khartoum.

"You're the right person to do the task," Jim said. "But as soon as you get that letter, you'll have only a few days to leave the Sudan."

With Jim's words, "You will have to leave the Sudan when Sadiq signs the letter," ringing in his ears, Cole returned to the prime minister's office another four or five times.

"There's no way I can get past his personal secretary to get the letter signed," Cole complained to me one afternoon. Finally, with the rumor mill suggesting Cole was on the verge of being declared *persona non grata*, Cole had an idea. We were seated on the angareb on our rooftop, and he told me: "Jasdip Singh, the World Bank representative, plays tennis with the Prime Minister every Friday . . ."

A few days later, Jasdip smiled as he took the envelope from his pocket and handed it to Cole. Reading the contents, Cole looked at Jasdip and breathed, "Sadiq has signed the letter!"

He then telephoned Jim Grant. "I have it!" he exclaimed.

The resulting network, Operation Lifeline Sudan, grew to be, as Vittachi says, "One of the great landmarks of humanitarian progress." But OLS was not without its critics.

"Grant is far too optimistic," other heads of UN agencies complained. Tarzie Vittachi reported in *Between the Guns* that the agency heads "were constantly being upstaged by Grant . . . envious of his successes and envious of the zealotry which they had never been able to produce either in themselves or in the work-ethic of the staffs."

Undeterred, Grant continued to defend what UNICEF was accomplishing in the Sudan and wrote a letter to the editor of *The Washington Post*, saying, "Perhaps the largest emergency relief effort ever mounted in so brief a period under such difficult circumstances is now under way. Both sides of a bitter civil war are working to make it possible and a vast array of international organizations, governments and private voluntary organizations are involved in this effort and doing an exemplary job. Yes, there are a myriad of problems, and deliveries are slower than needed, and yes, the effort probably will not succeed in saving the lives of all 100,000 people who would have died without it. But people are working together to resolve the problems and many tens of thousands of people will not die because this effort is being made."

Later, in a letter to *The Washington Post* on 19 May 1989, Representative Gary Ackerman criticized relief efforts, stating. "Overly optimistic strategies and assessments prevent us from squarely addressing the needs in southern Sudan."

Finally, Sudanese citizens aimed their frustration at relief workers in October 1990. "Why then do these odious relief organizations, who live as parasites on the resources of other people . . . and who thrust their gifts and assistance into the country, smuggle it into the ranks of the rebels and stab the Sudan in the back. *Allahu Akbar!* Pride be to Sudan!"

In agreement with the former Canadian Ambassador to the Sudan, who said, "The present Sudanese government has been the least cooperative of any I have ever dealt with," many NGOs wound down their operations in the hostile country. Others who stayed on in Khartoum found it difficult to obtain travel permits for fieldwork. CARE and Oxfam suffered attacks in the press and interference with hiring local staff. Their radios (essential for their rural work) were confiscated and their directors called in for intimidating interviews with security police.

Finally, Michael Priestly, the senior representative of the UN in the Sudan, told colleagues that a meeting he had with General Bashir on the question of the civil war and the obstruction of humanitarian aid was the worst he had ever experienced in his thirty-eight years with the United Nations.

Although there had been many rumors that the Sudanese government was upset with Cole for negotiating the letters which brought about Operation Lifeline Sudan, our leaving the country was amicable.

"I will not accept that you leave," Sadiq told Cole. "There are many rumors. People talk, but I would prefer that you stay. Let them complain. We view you as an asset and your commitment to the country is important for the people of Sudan. I trust you."

Replying that he had done what he could, Cole heard the Prime Minister say,

"There is only one way to justify your going. Will you get a promotion when you move to Dhaka?"

"Yes," Cole said.

Sadiq smiled and said, "Go well, and may God go with you."

About the Author

Marilyn S. Dodge coauthored *Sepia Prints: Memoirs of a Missionary in India* with her mother, Viola Wiebe. She and her husband Cole have three children, and are retired in Nairobi, Kenya.

CPSIA information can be obtained
at www.ICGtesting.com
Printed in the USA
LVHW092348040421
683424LV00005B/94

9 789966 757272